TinyOS Programming

Do you need to know how to write systems, services, and applications using the TinyOS operating system? Learn how to write nesC code and efficient applications with this indispensable guide to TinyOS programming.

Detailed examples show you how to write TinyOS code in full, from basic applications right up to new low-level systems and high-performance applications. Two leading figures in the development of TinyOS also explain the reasons behind many of the design decisions made and explain for the first time how nesC relates to and differs from other C dialects. Handy features such as a library of software design patterns, programming hints and tips, end-of-chapter exercises, and an appendix summarizing the basic application-level TinyOS APIs make this the ultimate guide to TinyOS for embedded systems programmers, developers, designers, and graduate students.

Philip Levis is Assistant Professor of Computer Science and Electrical Engineering at Stanford University. A Fellow of the Microsoft Research Faculty, he is also Chair of the TinyOS Core Working Group and a Member of the TinyOS Network Protocol (net2), Simulation (sim), and Documentation (doc) Working Groups.

David Gay joined Intel Research in Berkeley in 2001, where he has been a designer and the principal implementer of the nesC language, the C dialect used to implement the TinyOS sensor network operating system, and its applications. He has a diploma in Computer Science from the Swiss Federal Institute of Technology in Lausanne and a Ph.D. from the University of California, Berkeley.

TinyOS Programming

PHILIP LEVIS
Stanford University

and

DAVID GAY
Intel Research

CAMBRIDGE UNIVERSITY PRESS
Cambridge, New York, Melbourne, Madrid, Cape Town, Singapore, São Paulo, Delhi

Cambridge University Press
The Edinburgh Building, Cambridge CB2 8RU, UK

Published in the United States of America by Cambridge University Press, New York

www.cambridge.org
Information on this title: www.cambridge.org/9780521896061

© Cambridge University Press 2009

This publication is in copyright. Subject to statutory exception
and to the provisions of relevant collective licensing agreements,
no reproduction of any part may take place without
the written permission of Cambridge University Press.

First published 2009

Printed in the United Kingdom at the University Press, Cambridge

A catalogue record for this publication is available from the British Library

ISBN 978-0-521-89606-1 paperback

Cambridge University Press has no responsibility for the persistence or
accuracy of URLs for external or third-party internet websites referred to
in this publication, and does not guarantee that any content on such
websites is, or will remain, accurate or appropriate.

Contents

List of Code examples	*page* xi
Preface	xvii
Acknowledgements	xix
Programming hints, condensed	xxi

Part I TinyOS and nesC   1

1   Introduction   3
    1.1 Networked, embedded sensors   3
          1.1.1 Anatomy of a sensor node (mote)   4
    1.2 TinyOS   5
          1.2.1 What TinyOS provides   6
    1.3 Example application   7
    1.4 Compiling and installing applications   8
    1.5 The rest of this book   8

2   Names and program structure   10
    2.1 Hello World!   10
    2.2 Essential differences: components, interfaces, and wiring   13
    2.3 Wiring and callbacks   15
    2.4 Summary   16

Part II Basic programming   19

3   Components and interfaces   21
    3.1 Component signatures   21
          3.1.1 Visualizing components   22
          3.1.2 The "as" keyword and clustering interfaces   23
          3.1.3 Clustering interfaces   24
    3.2 Interfaces   24
          3.2.1 Generic interfaces   27
          3.2.2 Bidirectional interfaces   28

3.3	Component implementations	29
	3.3.1 Modules	30
	3.3.2 A basic configuration	31
	3.3.3 Module variables	32
	3.3.4 Generic components	33
3.4	Split-phase interfaces	34
	3.4.1 Read	36
	3.4.2 Send	36
3.5	Module memory allocation, avoiding recursion, and other details	36
	3.5.1 Memory ownership and split-phase calls	38
	3.5.2 Constants and saving memory	41
	3.5.3 Platform-independent types	42
	3.5.4 Global names	44
	3.5.5 nesC and the C preprocessor	46
	3.5.6 C libraries	47
3.6	Exercises	48

4 Configurations and wiring   49

4.1	Configurations	50
	4.1.1 The $->$ and $<-$ operators	51
	4.1.2 The = operator	52
	4.1.3 Namespace management	53
	4.1.4 Wiring rules	54
	4.1.5 Wiring shortcuts	56
4.2	Building abstractions	57
	4.2.1 Component naming	58
	4.2.2 Component initialization	59
4.3	Component layering	60
	4.3.1 Extensibility	61
	4.3.2 Hardware specificity	61
4.4	Multiple wirings	63
	4.4.1 Fan-in and fan-out	64
	4.4.2 Uses of multiple wiring	65
	4.4.3 Combine functions	66
4.5	Generics versus singletons	68
	4.5.1 Generic components, revisited	68
	4.5.2 Singleton components, revisited	70
4.6	Exercises	70

5 Execution model   71

5.1	Overview	71
5.2	Tasks	72
	5.2.1 Task timing	74

		5.2.2	Timing and event handlers	75
	5.3	Tasks and split-phase calls		75
		5.3.1	Hardware versus software	75
		5.3.2	Tasks and call loops	76
	5.4	Exercises		78

6 Applications 79

- 6.1 The basics: timing, LEDs, and booting — 79
  - 6.1.1 Deadline-based timing — 81
  - 6.1.2 Wiring AntiTheftC — 83
- 6.2 Sensing — 83
  - 6.2.1 Simple sampling — 84
  - 6.2.2 Sensor components — 85
  - 6.2.3 Sensor values, calibration — 86
  - 6.2.4 Stream sampling — 87
- 6.3 Single-hop networking — 89
  - 6.3.1 Sending packets — 90
  - 6.3.2 Receiving packets — 93
  - 6.3.3 Selecting a communication stack — 94
- 6.4 Multi-hop networking: collection, dissemination, and base stations — 95
  - 6.4.1 Collection — 96
  - 6.4.2 Dissemination — 97
  - 6.4.3 Wiring collection and dissemination — 97
  - 6.4.4 Base station for collection and dissemination — 98
- 6.5 Storage — 101
  - 6.5.1 Volumes — 102
  - 6.5.2 Configuration data — 103
  - 6.5.3 Block and Log storage — 105
- 6.6 Exercises — 111

7 Mote-PC communication 112

- 7.1 Basics — 112
  - 7.1.1 Serial communication stack — 113
- 7.2 Using mig — 114
  - 7.2.1 Sending and receiving mig-generated packets — 116
- 7.3 Using ncg — 118
- 7.4 Packet sources — 119
- 7.5 Example: simple reliable transmission — 120
  - 7.5.1 Reliable transmission protocol — 121
  - 7.5.2 Reliable transmission in Java — 121
  - 7.5.3 Reimplementing TestSerial — 125
- 7.6 Exercises — 125

Part III Advanced programming — 127

8 Advanced components — 129
- 8.1 Generic components review — 129
- 8.2 Writing generic modules — 131
  - 8.2.1 Type arguments — 132
  - 8.2.2 Abstract data types as generics — 133
  - 8.2.3 ADTs in TinyOS 1.x — 134
- 8.3 Parameterized interfaces — 135
  - 8.3.1 Parameterized interfaces and configurations — 137
  - 8.3.2 Parameterized interfaces and modules — 139
  - 8.3.3 Defaults — 141
- 8.4 Attributes — 142
- 8.5 Exercises — 144

9 Advanced wiring — 145
- 9.1 unique() and uniqueCount() — 145
  - 9.1.1 unique — 146
  - 9.1.2 uniqueCount — 147
  - 9.1.3 Example: HilTimerMilliC and VirtualizeTimerC — 147
- 9.2 Generic configurations — 150
  - 9.2.1 TimerMilliC — 150
  - 9.2.2 CC2420SpiC — 152
  - 9.2.3 AMSenderC — 156
  - 9.2.4 BlockStorageC — 160
- 9.3 Reusable component libraries — 162
- 9.4 Exercises — 165

10 Design patterns — 166
- 10.1 Behavioral: Dispatcher — 166
- 10.2 Structural: Service Instance — 170
- 10.3 Namespace: Keyspace — 174
- 10.4 Namespace: Keymap — 177
- 10.5 Structural: Placeholder — 180
- 10.6 Structural: Facade — 183
- 10.7 Behavioral: Decorator — 186
- 10.8 Behavioral: Adapter — 189

11 Concurrency — 192
- 11.1 Asynchronous code — 192
  - 11.1.1 The async keyword — 192
  - 11.1.2 The cost of async — 193
  - 11.1.3 Atomic statements and the atomic keyword — 195

		11.1.4 Managing state transitions	197

                11.1.4 Managing state transitions   197
                11.1.5 Example: CC2420ControlP   197
                11.1.6 Tasks, revisited   199
        11.2 Power locks   200
                11.2.1 Example lock need: link-layer acknowledgements   200
                11.2.2 Split-phase locks   201
                11.2.3 Lock internals   202
                11.2.4 Energy management   203
                11.2.5 Hardware configuration   204
                11.2.6 Example: MSP430 USART   204
                11.2.7 Power lock library   205
        11.3 Exercises   205

12 Device drivers and the hardware abstraction architecture (HAA)   206
        12.1 Portability and the hardware abstraction architecture   206
                12.1.1 Examples   208
                12.1.2 Portability   210
        12.2 Device drivers   210
                12.2.1 Access control   211
                12.2.2 Access control examples   212
                12.2.3 Power management   215
                12.2.4 Microcontroller power management   218
        12.3 Fitting in to the HAA   219

13 Advanced application: SoundLocalizer   221
        13.1 SoundLocalizer design   221
                13.1.1 Time synchronization   222
                13.1.2 Implementing SoundLocalizer in TinyOS   223
        13.2 SynchronizerC   225
        13.3 DetectorC   230
        13.4 MicrophoneC   233
        13.5 Wrap-up   237

Part IV Appendix and references   239

A TinyOS APIs   241
        A.1 Booting   241
        A.2 Communication   241
                A.2.1 Single-hop   242
                A.2.2 Multi-hop collection   243
                A.2.3 Multi-hop dissemination   244
                A.2.4 Binary reprogramming   245
        A.3 Time   245
        A.4 Sensing   245

A.5	Storage	246
A.6	Data structures	247
	A.6.1 BitVectorC	247
	A.6.2 QueueC	247
	A.6.3 BigQueueC	248
	A.6.4 PoolC	248
	A.6.5 StateC	249
A.7	Utilities	249
	A.7.1 Random numbers	249
	A.7.2 Leds	249
	A.7.3 Cyclic redundancy checks	250
	A.7.4 Printf	250
A.8	Low power	251

References   252

Index   254

Code examples

2.1	Powerup in C	*page*	10
2.2	PowerupC module in nesC		11
2.3	Simple nesC interfaces		11
2.4	PowerupAppC configuration in nesC		12
2.5	Powerup with blinking LED in C		15
2.6	Powerup with blinking LED in nesC (slightly simplified)		15
2.7	Powerup with blinking LED configuration (slightly simplified)		16
3.1	The signature and implementation blocks		21
3.2	Signatures of PowerupC and LedsC		22
3.3	MainC's signature		22
3.4	The LedsP module		23
3.5	PowerupC and an alternative signature		24
3.6	Interface declarations for Leds and Boot		25
3.7	The Init and Boot interfaces		25
3.8	Signatures of MainC and PowerupC		26
3.9	The Queue interface		27
3.10	Using a queue of 32-bit integers		27
3.11	Providing a 16-bit or a 32-bit queue		27
3.12	The Notify interface		28
3.13	UserButtonC		28
3.14	Simplified Timer interface showing three commands and one event		29
3.15	PowerupC module code		30
3.16	The module PowerupToggleC		30
3.17	The PowerupToggleAppC configuration		31
3.18	Example uses of the components keyword		31
3.19	The Get interface		32
3.20	A self-incrementing counter		32
3.21	Generic module SineSensorC and generic configuration TimerMilliC		33
3.22	Instantiating a generic component		34
3.23	Signature of BitVectorC		34
3.24	QueueC signature		34
3.25	The Read interface		36
3.26	The split-phase Send interface		36
3.27	The Send interface		38

3.28	The Receive interface	39
3.29	The signature of PoolC	41
3.30	CC2420 packet header	42
3.31	The dreaded "packed" attribute in the 1.x MintRoute library	43
3.32	The CC2420 header	44
3.33	TinyError.h, a typical nesC header file	45
3.34	Including a header file in a component	45
3.35	Indirectly including a header file	46
3.36	Fancy.nc: C preprocessor example	46
3.37	FancyModule.nc: C preprocessor pitfalls	47
3.38	Fancy.h: the reliable way to use C preprocessor symbols	47
3.39	Using a C library function	47
4.1	Signature of part of the CC1000 radio stack	49
4.2	The PowerupToggleAppC configuration revisited	51
4.3	C code generated from the PowerupToggleAppC configuration	51
4.4	The LedsC configuration	52
4.5	CC2420ReceiveC's use of the **as** keyword	53
4.6	Naming generic component instances	54
4.7	MainC and LedsP	55
4.8	Valid alternate of PowerupToggleAppC	55
4.9	Invalid alternate of PowerupToggleAppC	55
4.10	LedsC revisited	56
4.11	BlinkC signature	56
4.12	The RandomC configuration	57
4.13	The RandomMlcgC signature	58
4.14	Seed initialization in RandomMlcgP	59
4.15	ActiveMessageC for the CC2420	61
4.16	The signature of CC2420ActiveMessageC	62
4.17	Fan-out on CC2420TransmitC's Init	63
4.18	StdControl and SplitControl initialization interfaces	64
4.19	Why the metaphor of "wires" is only a metaphor	65
4.20	The combine function for error_t	66
4.21	Fan-out on SoftwareInit	67
4.22	Resulting code from fan-out on SoftwareInit	67
4.23	AMSenderC signature	68
4.24	RadioCountToLedsAppC	68
4.25	PoolC	69
4.26	Exposing a generic component instance as a singleton	70
5.1	The main TinyOS scheduling loop from SchedulerBasicP.nc	72
5.2	A troublesome implementation of a magnetometer sensor	76
5.3	Signal handler that can lead to an infinite loop	77
5.4	An improved implementation of FilterMagC	77
6.1	Anti-theft: simple flashing LED	80
6.2	The Leds interface	80

6.3	The Boot interface	81
6.4	The full Timer interface	81
6.5	WarningTimer.fired with drift problem fixed	82
6.6	Anti-Theft: application-level configuration	83
6.7	The Read interface	84
6.8	Anti-theft: detecting dark conditions	84
6.9	Anti-Theft: wiring to light sensor	86
6.10	ReadStream Interface	87
6.11	Anti-theft: detecting movement	88
6.12	The AMSend interface	90
6.13	Anti-Theft: reporting theft over the radio	91
6.14	The SplitControl interface	92
6.15	The Receive interface	93
6.16	Anti-Theft: changing settings	93
6.17	Serial vs Radio-based AM components	94
6.18	The Send interface	96
6.19	Anti-Theft: reporting theft over a collection tree	96
6.20	DisseminationValue interface	97
6.21	Anti-Theft: settings via a dissemination tree	97
6.22	The StdControl interface	97
6.23	The DisseminationUpdate interface	99
6.24	AntiTheft base station code: disseminating settings	99
6.25	The RootControl interface	100
6.26	AntiTheft base station code: reporting thefts	101
6.27	AntiTheft base station wiring	101
6.28	ConfigStorageC signature	102
6.29	Mount interface for storage volumes	103
6.30	ConfigStorage interface	103
6.31	Anti-Theft: reading settings at boot time	104
6.32	Anti-Theft: saving configuration data	105
6.33	BlockStorageC signature	106
6.34	The BlockWrite interface	106
6.35	Simultaneously sampling and storing to flash (most error checking omitted)	108
6.36	The BlockRead interface	108
6.37	LogStorageC signature	108
6.38	The LogWrite interface	109
6.39	The LogWrite interface	110
6.40	The LogRead interface	111
7.1	Serial AM Packet layout	113
7.2	TestSerial packet layout	114
7.3	Backing array methods	115
7.4	Sending packets with mig and MoteIF	117
7.5	Interface for handling received packets	117

7.6	Receiving packets with mig and MoteIF	117
7.7	Constants and packet layout for Oscillscope application	118
7.8	Class generated by ncg	119
7.9	Simplified code to save received samples	119
7.10	Reliable transmission protocol in Java – transmission	121
7.11	Reliable transmission protocol in Java – transmission	123
7.12	A reliable TestSerial.java	125
8.1	Instantiation within a generic configuration	130
8.2	The fictional component SystemServiceVectorC	131
8.3	QueueC excerpt	131
8.4	A generic constant sensor	132
8.5	Queue interface (repeated)	133
8.6	QueueC implementation	133
8.7	Representing an ADT through an interface in TinyOS 1.x	135
8.8	Timers without parameterized interfaces	135
8.9	Timers with a single interface	136
8.10	HilTimerMilliC signature	137
8.11	ActiveMessageC signature	138
8.12	Signature of TestAMC	138
8.13	Wiring TestAMC to ActiveMessageC	138
8.14	A possible module underneath ActiveMessageC	139
8.15	Parameterized interface syntax	140
8.16	Dispatching on a parameterized interface	140
8.17	How active message implementations decide on whether to signal to Receive or Snoop	140
8.18	Defining a parameter	141
8.19	Wiring full parameterized interface sets	141
8.20	Default events in an active message implementation	142
8.21	nesC attributes	143
9.1	Partial HilTimerMilliC signature	146
9.2	VirtualizeTimerC	148
9.3	Instantiating VirtualizeTimerC	148
9.4	VirtualizeTimerC state allocation	149
9.5	The TimerMilliC generic configuration	151
9.6	TimerMilliP auto-wires HilTimerMilliC to Main.SoftwareInit	151
9.7	The Blink application	151
9.8	The full module-to-module wiring chain in Blink (BlinkC to VirtualizeTimerC)	152
9.9	CC2420SpiC	153
9.10	CC2420SpiP	154
9.11	CC2420SpiC mappings to CC2420SpiP	154
9.12	The strobe implementation	155
9.13	The AMSenderC generic configuration	158
9.14	AMSendQueueEntryP	159

List of code examples

9.15	AMQueueP	159
9.16	AMSendQueueImplP pseudocode	160
9.17	BlockStorageC	161
9.18	The full code of HilTimerMilliC	163
9.19	VirtualizeTimerC virtualizes a single timer	164
10.1	AMReceiverC	169
10.2	VirtualizeTimerC	172
10.3	Telos ActiveMessageC	181
10.4	The Matchbox facade	184
10.5	The CC2420CsmaC uses a Facade	185
10.6	AlarmToTimerC implementation	190
11.1	The Send interface	192
11.2	The Leds interface	193
11.3	Toggling a state variable	193
11.4	A call sequence that could corrupt a variable	194
11.5	State transition that is not async-safe	194
11.6	Incrementing with an atomic statement	195
11.7	Incrementing with two independent atomic statements	195
11.8	The first step of starting the CC2420 radio	198
11.9	The handler that the first step of starting the CC2420 is complete	198
11.10	The handler that the second step of starting the CC2420 is complete	198
11.11	The handler that the third step of starting the CC2420 radio is complete	199
11.12	State transition so components can send and receive packets	199
11.13	The Resource interface	201
11.14	Msp430Spi0C signature	202
11.15	Msp320Adc12ClientC signature	202
11.16	The ResourceDefaultOwner interface	203
11.17	The ResourceConfigure interface	204
12.1	ActiveMessageC signature	212
12.2	Arbitration in Stm25pSectorC	215
12.3	McuSleepC: platform-specific sleep code	218
13.1	SynchronizerC: time synchronization for SoundLocalizer	225
13.2	The Counter interface	226
13.3	DetectorC: loud sound detection for SoundLocalizer	231
13.4	The Alarm interface	231
13.5	Atm128AdcSingle: low-level single-sample ATmega128 A/D converter interface	232
13.6	The GeneralIO digital I/O pin interface	235
13.7	The I2CPacket interface for bus masters	236

Preface

This book provides an in-depth introduction to writing nesC code for the TinyOS 2.0 operating system. While it goes into greater depth than the TinyOS tutorials on this subject, there are several topics that are outside its scope, such as the structure and implementation of radio stacks or existing TinyOS libraries. It focuses on how to write nesC code, and explains the concepts and reasons behind many of the nesC and TinyOS design decisions. If you are interested in a brief introduction to TinyOS programming, then you should probably start with the tutorials. If you're interested in details on particular TinyOS subsystems you should probably consult TEPs (TinyOS Enhancement Proposals), which detail the corresponding design considerations, interfaces, and components. Both of these can be found in the `doc/html` directory of a TinyOS distribution.

While some of the contents of this book are useful for 1.x versions of TinyOS, they do have several differences from TinyOS 2.0 which can lead to different programming practices. If in doubt, referring to the TEP on the subject is probably the best bet, as TEPs often discuss in detail the differences between 1.x and 2.0.

For someone who has experience with C or C++, writing simple nesC programs is fairly straightforward: all you need to do is implement one or two modules and wire them together. The difficulty (and intellectual challenge) comes when building larger applications. The code inside TinyOS modules is fairly analogous to C coding, but configurations – which stitch together components – are not.

This book is a first attempt to explain how nesC relates to and differs from other C dialects, stepping through how the differences lead to very different coding styles and approaches. As a starting point, this book assumes that

1. you know C, C++, or Java reasonably well, understand pointers and that
2. you have taken an undergraduate level operating systems class (or equivalent) and know about concurrency, interrupts, and preemption.

Of course, this book is as much a description of nesC as it is an argument for a particular way of using the language to achieve software engineering goals. In this respect, it is the product of thousands of hours of work by many people, as they learned and explored the use of the language. In particular, Cory Sharp, Kevin Klues, and Vlado Handziski have always pushed the boundaries of nesC programming in order to better understand which practices lead to the simplest, most efficient, and robust code. In particular, Chapter 10 is an edited version of a paper we wrote together, while using structs as a compile-time checking mechanism in interfaces (as Timer does) is an approach invented by Cory.

This book is divided into four parts. The first part, Chapters 1–2, gives a high-level overview of TinyOS and the nesC language. The second part, Chapters 3–7 goes into nesC and TinyOS at a level sufficient for writing applications. The third part, Chapters 8–13 goes into more advanced TinyOS and nesC programming, as is sometimes needed when writing new low-level systems or high performance applications. The book ends with an appendix summarizing the basic application-level TinyOS APIs.

Acknowledgements

We'd like to thank several people for their contributions to this book. First is Mike Horton, of Crossbow, Inc., who first proposed writing it. Second is Pablo Guerrero, who gave detailed comments and corrections. Third is Joe Polastre of Moteiv, who gave valuable feedback on how to better introduce generic components. Fourth, we'd like to thank Phil's father, who although he doesn't program, read the entire thing! Fifth, John Regehr, Ben Greenstein and David Culler provided valuable feedback on this expanded edition. Last but not least, we would like to thank the TinyOS community and its developers. Many of the concepts in this book – power locks, tree routing, and interface type checking – are the work and ideas of others, which we merely present.

Chapter 10 of this book is based on: Software design patterns for TinyOS, in ACM Transactions on Embedded Computing Systems (TECS), Volume 6, Issue 4 (September 2007), ©ACM, 2007. http://doi.acm.org/10.1145/1274858.1274860

Programming hints, condensed

Programming Hint 1 Use the "as" keyword liberally. (page 24)

Programming Hint 2 Never write recursive functions within a module. In combination with the TinyOS coding conventions, this guarantees that all programs have bounded stack usage. (page 38)

Programming Hint 3 Never use malloc and free. Allocate all state in components. If your application requirements necessitate a dynamic memory pool, encapsulate it in a component and try to limit the set of users. (page 38)

Programming Hint 4 When possible, avoid passing pointers across interfaces; when this cannot be avoided only one component should be able to modify a pointer's data at any time. (page 39)

Programming Hint 5 Conserve memory by using enums rather than const variables for integer constants, and don't declare variables with an enum type. (page 42)

Programming Hint 6 Never, ever use the "packed" attribute in portable code. (page 43)

Programming Hint 7 Use platform-independent types when defining message structures. (page 44)

Programming Hint 8 If you have to perform significant computation on a platform-independent type or access it many (hundreds or more) times, temporarily copy it to a native type. (page 44)

Programming Hint 9 Interfaces should #include the header files for the types they use. (page 46)

Programming Hint 10 Always #define a preprocessor symbol in a header file. Use #include to load the header file in all components and interfaces that use the symbol. (page 47)

Programming Hint 11 If a component is a usable abstraction by itself, its name should end with C. If it is intended to be an internal and private part of a larger abstraction, its name should end with P. Never wire to P components from outside your package (directory). (page 58)

Programming Hint 12 Auto-wire Init to MainC in the top-level configuration of a software abstraction. (page 60)

Programming Hint 13 When using layered abstractions, components should not wire across multiple abstraction layers: they should wire to a single layer. (page 63)

Programming Hint 14 Never ignore combine warnings. (page 68)

Programming Hint 15 Keep tasks short. (page 74)

Programming Hint 16 If an event handler needs to make possibly long-executing command calls, post a task to make the calls. (page 75)

Programming Hint 17 Don't signal events from commands – the command should post a task that signals the event. (page 77)

Programming Hint 18 Use a parameterized interface when you need to distinguish callers or when you have a compile-time constant parameter. (page 141)

Programming Hint 19 If a component depends on unique, then #define the string to use in a header file, to prevent bugs due to string typos. (page 149)

Programming Hint 20 Whenever writing a module, consider making it more general-purpose and generic. In most cases, modules must be wrapped by configurations to be useful, so singleton modules have few advantages. (page 165)

Programming Hint 21 Keep code synchronous when you can. Code should be async only if its timing is very important or if it might be used by something whose timing is important. (page 195)

Programming Hint 22 Keep atomic statements short, and have as few of them as possible. Be careful about calling out to other components from within an atomic statement. (page 199)

Part I

TinyOS and nesC

1 Introduction

This book is about writing TinyOS systems and applications in the nesC language. This chapter gives a brief overview of TinyOS and its intended uses. TinyOS is an open-source project which a large number of research universities and companies contribute to. The main TinyOS website, www.tinyos.net, has instructions for downloading and installing the TinyOS programming environment. The website has a great deal of useful information which this book doesn't cover, such as common hardware platforms and how to install code on a node.

1.1 Networked, embedded sensors

TinyOS is designed to run on small, wireless sensors. Networks of these sensors have the potential to revolutionize a wide range of disciplines, fields, and technologies. Recent example uses of these devices include:

Golden Gate Bridge safety High-speed accelerometers collect synchonized data on the movement of and oscillations within the structure of San Francisco's Golden Gate Bridge. This data allows the maintainers of the bridge to easily observe the structural health of the bridge in response to events such as high winds or traffic, as well as quickly assess possible damage after an earthquake [10]. Being wireless avoids the need for installing and maintaining miles of wires.

Volcanic monitoring Accelerometers and microphones observe seismic events on the Reventador and Tungurahua volcanoes in Ecuador. Nodes locally compare when they observe events to determine their location, and report aggregate data to a camp several kilometers away using a long-range wireless link. Small, wireless nodes allow geologists and geophysicists to install dense, remote scientific instruments [30], obtaining data that answers other questions about unapproachable environments.

Data center provisioning Data centers and enterprise computing systems require huge amounts of energy, to the point at which they are placed in regions that have low power costs. Approximately 50% of the energy in these systems goes into cooling, in part due to highly conservative cooling systems. By installing wireless sensors across machine racks, the data center can automatically sense what areas need cooling and can adjust which computers do work and generate heat [19]. Dynamically adapting

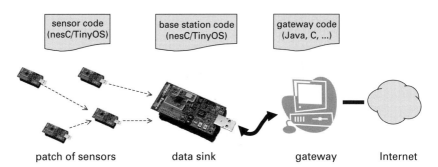

Figure 1.1 A typical sensor network architecture. Patches of ultra-low power sensors, running nesC/TinyOS, communicate to gateway nodes through data sinks. These gateways connect to the larger Internet.

these factors can greatly reduce power consumption, making the IT infrastructure more efficient and reducing environmental impact.

While these three application domains are only a small slice of where networks of sensors are used, they show the key differences between these networks and most other computing systems. First, these "sensor networks" need to operate unattended for long periods of time. Second, they gather data from and respond to an unpredictable environment. Finally, for reasons of cost, deployment simplicity, and robustness, they are wireless. Together, these three issues – longevity, embedment, and wireless communication – cause sensor networks to use different approaches than traditional, wired, and human-centric or machine-centric systems.

The sheer diversity of sensor network applications means that there are many network architectures, but a dominant portion of deployments tend to follow a common one, shown in Figure 1.1 [21, 26, 30] of ultra-low power sensors self-organized to form an ad-hoc routing network to one or more data sink nodes. These sensor sinks are attached to gateways, which are typically a few orders of magnitude more powerful than the sensors: gateways run an embedded form of Linux, Windows, or other multitasking operating system. Gateways have an Internet connection, either through a cell phone network, long-distance wireless, or even just wired Ethernet.

Energy concerns dominate sensor hardware and software design. These nodes need to be wireless, small, low-cost, and operate unattended for long periods. While it is often possible to provide large power resources, such as large solar panels, periodic battery replacement, or wall power, to small numbers of gateways, doing so to every one of hundreds of sensors is infeasible.

1.1.1 Anatomy of a sensor node (mote)

Since energy consumption determines sensor node lifetime, sensor nodes, commonly referred to as *motes*, tend to have very limited computational and communication resources. Instead of a full-fledged 32-bit or 64-bit CPU with megabytes or gigabytes of RAM, they have 8-bit or 16-bit microcontrollers with a few kilobytes of RAM. Rather than gigahertz, these microcontrollers run at 1–10 megahertz. Their low-power radios

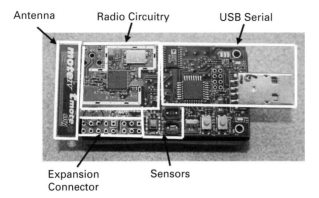

Figure 1.2 A Telos sensor produced by Moteiv. The top of the node has the radio, sensors, and circuitry for the USB connector. The bottom, not shown, has the processor and flash storage chip. The antenna is part of the printed circuit board (PCB).

can send tens to hundreds of kilobits per second (kbps), rather than 802.11's tens of megabits. As a result, software needs to be very efficient, both in terms of CPU cycles and in terms of memory use.

Figure 1.2 shows a sample node platform, the Telos, which is designed for easy experimentation and low-power operation. It has a TI MSP430 16-bit microcontroller with 10 kB of RAM and 48 kB of flash program memory. Its radio, a TI CC2420 which follows the IEEE 802.15.4 standard, can send up to 250 kbps. In terms of power, the radio dominates the system: on a pair of AA batteries, a Telos can have the radio on for about four days. Lasting longer than four days requires keeping the node in a deep sleep state most of the time, waking only when necessary, and sleeping as soon as possible.

The other mote discussed in this book, the micaz from Crossbow Technology is similar: it has an Atmel ATmega128 8-bit microcontroller with 4 kB of RAM, 128 kB of flash program memory, uses the same CC2420 radio chip, also runs off a pair of AA batteries and has a similar power consumption profile.

Networks, once deployed, gather data uninterrupted for weeks, months, or years. As the placement of sensors is very application-specific, it is rare for networks to need to support multiple concurrent applications, or even require more than the occasional reprogramming. Therefore, unlike general-purpose computing systems, which emphasize run-time flexibility and composability, sensor network systems tend to be highly optimized. Often, the sensor suite itself is selected for the specific application: volcanic monitoring uses accelerometers and microphones, while data center provisioning uses temperature sensors.

1.2 TinyOS

TinyOS is a lightweight operating system specifically designed for low-power wireless sensors. TinyOS differs from most other operating systems in that its design focuses on ultra low-power operation. Rather than a full-fledged processor, TinyOS is designed for the small, low-power microcontrollers motes have. Furthermore, TinyOS has very aggressive systems and mechanisms for saving power.

TinyOS makes building sensor network applications easier. It provides a set of important services and abstractions, such as sensing, communication, storage, and timers. It defines a concurrent execution model, so developers can build applications out of reusable services and components without having to worry about unforeseen interactions. TinyOS runs on over a dozen generic platforms, most of which easily support adding new sensors. Furthermore, TinyOS's structure makes it reasonably easy to port to new platforms.

TinyOS applications and systems, as well as the OS itself, are written in the nesC language. nesC is a C dialect with features to reduce RAM and code size, enable significant optimizations, and help prevent low-level bugs like race conditions. Chapter 2 goes into the details on how nesC differs significantly from other C-like languages, and most of this book is about how to best use those features to write robust, efficient code.

1.2.1 What TinyOS provides

At a high level, TinyOS provides three things to make writing systems and applications easier:

- a component model, which defines how you write small, reusable pieces of code and compose them into larger abstractions;
- a concurrent execution model, which defines how components interleave their computations as well as how interrupt and non-interrupt code interact;
- application programming interfaces (APIs), services, component libraries and an overall component structure that simplify writing new applications and services.

The component model is grounded in nesC. It allows you to write pieces of reusable code which explicitly declare their dependencies. For example, a generic user button component that tells you when a button is pressed sits on top of an interrupt handler. The component model allows the button implementation to be independent of which interrupt that is – e.g. so it can be used on many different hardware platforms – without requiring complex callbacks or magic function naming conventions. Chapter 2 and Chapter 3 describe the basic component model.

The concurrent execution model enables TinyOS to support many components needing to act at the same time while requiring little RAM. First, every I/O call in TinyOS is *split-phase*: rather than block until completion, a request returns immediately and the caller gets a callback when the I/O completes. Since the stack isn't tied up waiting for I/O calls to complete, TinyOS only needs one stack, and doesn't have threads. Instead, Chapter 5 introduces *tasks*, which are lightweight deferred procedure calls. Any component can post a task, which TinyOS will run at some later time. Because low-power devices must spend most of their time asleep, they have low CPU utilization and so in practice tasks tend to run very soon after they are posted (within a few milliseconds). Furthermore, because tasks can't preempt each other, task code doesn't need to worry about data races. Low-level interrupt code (discussed in the advanced concurrency chapter, Chapter 11) can have race conditions, of course: nesC detects possible data races at compile-time and warns you.

Finally, TinyOS itself has a set of APIs for common functionality, such as sending packets, reading sensors, and responding to events. Uses of these are sprinkled throughout the entire book, and presented in more detail in Chapter 6 and Appendix 1. In addition to programming interfaces, TinyOS also provides a component structure and component libraries. For example, Chapter 12 describes TinyOS's Hardware Abstraction Architecture (HAA), which defines how to build up from low-level hardware (e.g. a radio chip) to a hardware-independent abstraction (e.g. sending packets). Part of this component structure includes resource locks, covered in Chapter 11, which enable automatic low-power operation, as well as the component libraries that simplify writing such locks.

TinyOS itself is continually evolving. Within the TinyOS community, "Working Groups" form to tackle engineering and design issues within the OS, improving existing services and adding new ones. This book is therefore really a snapshot of the OS in time. As Chapter 12 discusses and Appendix 1 presents, TinyOS has a set of standard, stable APIs for core abstractions, but this set is always expanding as new hardware and applications emerge. The best way to stay up to date with TinyOS is to check its web page www.tinyos.net and participate in its mailing lists. The website also covers advanced TinyOS and nesC features which are well beyond the scope of this book, including binary components, over-the-air reprogramming services, debugging tools, and a nesC reference manual.

1.3 Example application

To better understand the unique challenges faced by sensor networks, we walk through a basic data-collection application. Nodes running this application periodically wake up, sample some sensors, and send the data through an ad hoc collection tree to a data sink (as in Figure 1.1). As the network must last for a year, nodes spend 99% of their time in a deep sleep state.

In terms of energy, the radio is by far the most expensive part of the node. Lasting a year requires telling the radio to be in a low power state. Low power radio implementation techniques are beyond the scope of this book, but the practical upshot is that packet transmissions have higher latency. [23]

Figure 1.3 shows the four TinyOS APIs the application uses: low power settings for the radio, a timer, sensors, and a data collection routing layer. When TinyOS tells the application that the node has booted, the application code configures the power settings on the radio and starts a periodic timer. Every few minutes, this timer fires and the application code samples its sensors. It puts these sensor values into a packet and calls the routing layer to send the packet to a data sink. In practice, applications tend to be more complex than this simple example. For example, they include additional services such as a management layer which allows an administrator to reconfigure parameters and inspect the state of the network, as well as over-the-air programming so the network

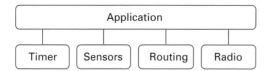

Figure 1.3 Example application architecture. Application code uses a timer to act periodically, sensors to collect data, and a routing layer to deliver data to a sink.

can be reprogrammed without needing to collect all of the nodes. However, these four abstractions – power control, timers, sensors, and data collection – encompass the entire datapath of the application.

1.4 Compiling and installing applications

You can download the latest TinyOS distribution, the nesC compiler, and other tools at www.tinyos.net. Setting up your programming environment is outside the scope of this book; the TinyOS website has step-by-step tutorials to get you started. One part of TinyOS is an extensive build system for compiling applications. Generally, to compile a program for a sensor platform, one types make <platform>, e.g. make telosb. This compiles a binary. To install that binary on a node, you plug the node into your PC using a USB or serial connection, and type make <platform> install. The tutorials go into compilation and installation options in detail.

1.5 The rest of this book

The rest of this book goes into how to program in nesC and write TinyOS applications. It is divided into three parts. The first is a short introduction to the major programming concepts of nesC. The second part addresses basic application programming using standard TinyOS APIs. The third part digs a little deeper, and looks into how those TinyOS APIs are implemented. For example, the third part describes how TinyOS abstracts hardware, so you can write a driver for a new sensor.

Chapter by chapter, the book is structured as follows:

- **Chapter 1** is this chapter.
- **Chapter 2** describes the major way that nesC breaks from C and C-like languages: how programs are built out of components, and how components and interfaces help manage programs' namespaces.
- **Chapter 3** presents components and how they interact via interfaces.
- **Chapter 4** goes into greater detail into configurations, components which connect other components together.
- **Chapter 5** covers the basic TinyOS execution model and gives guidance on how and when to use tasks.

1.5 The rest of this book

- **Chapter 6** takes the material from the prior three chapters and brings it together into an example of writing a fully-fledged application that sends an alarm when a node observes a change in its environment. In the process, it covers the major TinyOS APIs (timing, sensing, communication, and storage).
- **Chapter 7** details the PC-side tools for communicating with nodes connected over the serial port, and covers the TinyOS serial stack and packet formats.
- **Chapter 8** introduces more advanced component topics, such as parameterized interfaces and attributes. While applications typically don't use these mechanisms, they are indispensable when writing reusable libraries and systems.
- **Chapter 9** goes into wiring parameterized interfaces, which form the basis for most reusable systems. After describing the basic mechanisms for managing interface keys, it goes through four examples of increasing complexity.
- **Chapter 10** presents eight common TinyOS design patterns: commonly useful and powerful ways of organizing components.
- **Chapter 11** concludes the advanced programming topics by covering concurrency. It describes asynchronous code, the TinyOS task model, and power locks.
- **Chapter 12** describes the Hardware Abstraction Architecture (HAA), the three-level hierarchy TinyOS uses to raise low-level hardware abstractions to hardware-independent, application-level services.
- **Chapter 13** goes step-by-step through an advanced application that identifies the location of an event based on when nodes sense it. It uses lower-layer interfaces to implement time synchronization and high-frequency sampling.
- **Appendix** gives a concise overview of major TinyOS application interfaces.

Throughout the book, you will find *programming hints*: general best-use practices that we've distilled from the many users of TinyOS and nesC. These are all listed on the sheet at the front of the book.

Finally, the complete source code for example applications presented in this book (in Chapters 6, 7 and 13) is available in TinyOS's contributed code directory, under the name "TinyOS Programming" – see www.tinyos.net for details.

2 Names and program structure

Program structure is the most essential and obvious difference between C and nesC. C programs are composed of variables, types, and functions defined in files that are compiled separately and then linked together. nesC programs are built out of components that are connected ("wired") by explicit program statements; the nesC compiler connects and compiles these components as a single unit. To illustrate and explain these differences in how programs are built, we compare and contrast C and nesC implementations of two very simple "hello world"-like mote applications, Powerup (boot and turn on a LED) and Blink (boot and repeatedly blink a LED).

2.1   Hello World!

The closest mote equivalent to the classic "Hello World!" program is the "Powerup" application that simply turns on one of the motes LEDs at boot, then goes to sleep.

A C implementation of Powerup is fairly simple:

```c
#include "mote.h"

int main()
{
  mote_init();
  led0_on();
  sleep();
}
```

Listing 2.1 Powerup in C

The Powerup application is compiled and linked with a "mote" library which provides functions to perform hardware initialization (mote_init), LED control (led0_on) and put the mote in to a low-power sleep mode (sleep). The "mote.h" header file simply provides declarations of these and other basic functions. The usual C main function is called automatically when the mote boots.[1]

---

[1] The C compiler, library, and linker typically arrange for this by setting the mote's hardware reset vector to point to a piece of assembly code that sets up a C environment, then calls main.

The nesC implementation of Powerup is split into two parts. The first, the PowerupC *module*, contains the executable logic of Powerup (what there is of it ...):

```
module PowerupC {
  uses interface Boot;
  uses interface Leds;
}
implementation {
  event void Boot.booted() {
    call Leds.led0On();
  }
}
```

Listing 2.2 PowerupC module in nesC

This code says that PowerupC interacts with the rest of the system via two *interfaces*, Boot and Leds, and provides an implementation for the booted *event* of the Boot interface that calls the led0On[2] *command* of the Leds interface. Comparing with the C code, we can see that the booted event implementation takes the place of the main function, and the call to the led0On command the place of the call to the led0_on library function.

This code shows two of the major differences between nesC and C: where C programs are composed of functions, nesC programs are built out of *components* that implement a particular service (in the case of PowerupC, turning a LED on at boot-time). Furthermore, C functions typically interact by calling each other directly, while the interactions between components are specified by interfaces: the interface's *user* makes requests (*calls commands*) on the interface's *provider*, the provider makes callbacks (*signals events*) to the interface's user. Commands and events themselves are like regular functions (they can contain arbitrary C code); calling a command or signaling an event is just a function call. PowerupC is a user of both Boot and Leds; the booted event is a callback signaled when the system boots, while the led0On is a command requesting that LED 0 be turned on.

nesC interfaces are similar to Java interfaces, with the addition of a `command` or `event` keyword to distinguish requests from callbacks:

```
interface Boot} {
  event void} booted();
}

interface Leds {
  command void led0On();
  command void led0Off();
```

---

[2] LEDs are numbered in TinyOS, as different platforms have different color LEDs.

```
        command void led0Toggle();
        ...
}
```

Listing 2.3 Simple nesC interfaces

The second part of Powerup, the PowerupAppC *configuration*, specifies how PowerupC is connected to TinyOS's services:

```
configuration PowerupAppC { }
implementation {
  components MainC, LedsC, PowerupC;

  MainC.Boot  -> PowerupC.Boot;
  PowerupC.Leds -> LedsC.Leds;
}
```

Listing 2.4 PowerupAppC configuration in nesC

This says that the PowerupAppC application is built out of three *components* (modules or configurations), MainC (system boot), LedsC (LED control), and PowerupC (our powerup module). PowerupAppC explicitly specifies the connections (or *wiring*) between the interfaces provided and used by these components. When MainC has finished booting the system it *signals* the booted event of its Boot interface, which is connected by the wiring in PowerupAppC to the booted event in PowerupC. This event then calls the led0On command of its Leds interface, which is again connected (*wired*) by PowerupAppC to the Leds interface provided by LedsC. Thus the call turns on LED 0. The resulting component diagram is shown in Figure 2.1 – this diagram was generated automatically from PowerupAppC by nesdoc, nesC's documentation generation tool.

PowerupAppC illustrates the third major difference between C and nesC: wiring makes the connections expressed by linking the C version of Powerup with its "mote" library explicit. In the C version, Powerup calls a global function named led0_on which is connected to whatever library provides a function with the same name; if two libraries

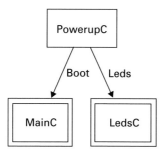

Figure 2.1 Wiring Diagram for Powerup application

2.2 Essential differences: components, interfaces, and wiring

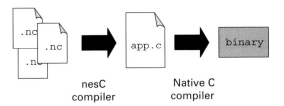

Figure 2.2 The nesC compilation model. The nesC compiler loads and reads in nesC components, which it compiles to a C file. This C file is passed to a native C compiler, which generates a mote binary.

provide such a function then (typically) the first one named on the linker command line "wins." Using a nesC configuration, the programmer instead explicitly selects which component's implementation of the function to use.

The nesC compiler can take advantage of this explicit wiring to build highly optimized binaries. Current implementations of the nesC compiler (nesc1) take nesC files describing components as input and output a C file. The C file is passed to a native C compiler that can compile to the desired microcontroller or processor. Figure 2.2 shows this process. The nesC compiler carefully constructs the generated C file to maximize the optimization abilities of the C compiler. For example, since it is given a single file, the C compiler can freely optimize across call boundaries, inlining code whenever needed. The nesC compiler also prunes dead code which is never called and variables which are never accessed: since there is no dynamic linking in nesC, it has a complete picture of the application call graph. This speeds the C compilation and reduces program size in terms of both RAM and code.

2.2 Essential differences: components, interfaces, and wiring

The three essential differences between C and nesC – components, interfaces, and wiring – all relate to naming and organizing a program's elements (variables, functions, types, etc). In C, programs are broken into separate files which are connected via a *global namespace*: a symbol *X declared* in one file is connected by the linker to a symbol *X defined* in another file. For instance, if `file1.c` contains:

```
extern void g(void);  /* declaration of g */

int main()  /* definition of main */
{
  g(); g();
}
```

and `file2.c` contains:

```
void g(void)
{
  printf("hello world!");
}
```

then compiling and linking `file1.c` and `file2.c` connects the calls to `g()` in main to the definition of g in `file2.c`. The resulting program prints "hello world!" twice.

Organizing symbols in a global namespace can be tricky. C programmers use a number of techniques to simplify this task, including header files and naming conventions. Header files group declarations so they can be used in a number of files without having to retype them, e.g. a header file `file1.h` for `file1.c` would normally contain:

```
#ifndef FILE1_H
#define FILE1_H
extern void g(void);  /* declaration of g */
#endif
```

Naming conventions are designed to avoid having two different symbols with the same name. For instance, types are often suffixed with _t guaranteeing that a type and function won't have the same name. Some libraries use a common prefix for all their symbols, e.g. Gtk and gtk_ for the GTK+ graphical toolkit. Such prefixes remind users that functions are related and avoid accidental name collisions with other libraries, but make programs more verbose.

nesC's components provide a more systematic approach for organizing a program's elements. A component (module or configuration) groups related functionality (a timer, a sensor, system boot) into a single unit, in a way that is very similar to a class in an object-oriented language. For instance, TinyOS represents its system services as separate components such as LedsC (LED control, seen above), ActiveMessageC (sending and receiving radio messages), etc. Only the service (component) name is global, the service's operations are named in a per-component scope: ActiveMessageC.SplitControl starts and stops the radio, ActiveMessageC.AMSend sends a radio message, etc.

Interfaces bring further structure to components: components are normally specified in terms of the set of interfaces (Leds, Boot, SplitControl, AMSend) that they provide and use, rather than directly in terms of the actual operations. Interfaces simplify and clarify code because, in practice, interactions between components follow standard patterns: many components want to control LEDs or send radio messages, many services need to be started or stopped, etc. Encouraging programmers to express their components in terms of common interfaces also promotes code reuse: expressing your new network protocol in terms of the AMSend message transmission interface means it can be used with existing applications, using AMSend in your application means that it can be used with any existing or future network protocol.

Rather than connect declarations to definitions with the same name, nesC programs use wiring to specify how components interact: PowerupAppC wired PowerupC's Leds interface to that provided by the LedsC component, but a two-line change could switch that wiring to the `NoLedsC` component (which just does nothing):

```
components PowerupC, NoLedsC;
PowerupC.LedsC -> NoLedsC.Leds;
```

without affecting any other parts of the program that wish to use LedsC. In C, one could replace the "mote" library used by Powerup by a version where the LED functions did nothing, but that change would affect all LED users, not just Powerup.

2.3 Wiring and callbacks

Leaving the component connection decisions to the programmer does more than just simplify switching between multiple service implementations. It also provides an efficient mechanism for supporting callbacks, as we show through the example of timers. TinyOS provides a variable number of periodic or deadline timers; associated with each timer is a callback to a function that is executed each time the timer fires. We first look at how such timers would be expressed in C, by modifying Powerup to blink LED 0 at 2 Hz rather than turn it on once and for all:

```c
#include "mote.h"

timer_t mytimer;

void blink_timer_fired(void)
{
  leds0_toggle();
}

int main()
{
  mote_init();
  timer_start_periodic(&mytimer, 250, blink_timer_fired);
  sleep();
}
```

Listing 2.5 Powerup with blinking LED in C

In this example, the Blink application declares a global mytimer variable to hold timer state, and calls timer_start_periodic to set up a periodic 250 ms timer. Every time the timer fires, the timer implementation performs a callback to the blink_timer_fired function specified when the timer was set up. This function simply calls a library function that toggles LED 0 on or off.

The nesC version of Blink is similar to the C version, but uses interfaces and wiring to specify the connection between the timer and the application:

```nesc
module BlinkC {
  uses interface Boot;
  uses interface Timer;
  uses interface Leds;
}
implementation {
  event void Boot.booted() {
    call Timer.startPeriodic(250);
  }

  event void Timer.fired() {
```

```
    call Leds.led0Toggle();
  }
}
```

Listing 2.6 Powerup with blinking LED in nesC (slightly simplified)

The BlinkC module starts the periodic 250 ms timer when it boots. The connection between the startPeriodic command that starts the timer and the fired event which blinks the LED is implicitly specified by having the command and event in the same interface:

```
interface Timer {
  command void startPeriodic(uint32_t interval);
  event void fired();
  ...
}
```

Finally, this Timer must be connected to a component that provides an actual timer. BlinkAppC wires BlinkC.Timer to a newly allocated timer MyTimer:

```
configuration BlinkAppC { }
implementation {
  components MainC, LedsC, new TimerC() as MyTimer, BlinkC;

  BlinkC.Boot  -> MainC.Boot;
  BlinkC.Leds  -> LedsC.Leds;
  BlinkC.Timer -> MyTimer.Timer;
}
```

Listing 2.7 Powerup with blinking LED configuration (slightly simplified)

In the C version the callback from the timer to the application is a run-time argument to the timer_start_periodic function. The timer implementation stores this function pointer in the mytimer variable that holds the timer's state, and performs an indirect function call each time the timer fires. Conversely, in the nesC version, the connection between the timer and the Blink application is specified at compile-time in BlinkAppC. This avoids the need to store a function pointer (saving precious RAM), and allows the nesC compiler to perform optimizations (in particular, inlining) across callbacks.

2.4 Summary

Table 2.1 summarizes the difference in how programs are structured in C, C++ and nesC. In C, the typical high-level programming unit is the file, with an associated header file that specified and documents the file's behavior. The linker builds applications out of files by matching global names; where this is not sufficient to express program structure (e.g. for callbacks), the programmer can use function pointers to delay the decision of which function is called at what point.

Table 2.1. Program Structure in C, C++ and nesC

structural element	C	C++	nesC
program unit	file	class	component
unit specification	header file	class declaration	component specification
specification pattern	–	abstract class	interface
unit composition	name matching	name matching	wiring
delayed composition	function pointer	virtual method	wiring

C++ provides explicit language mechanisms for structuring programs: classes are typically used to group related functionality, and programs are built out of interacting objects (class instances). An abstract class can be used to define common class specification patterns (like sending a message); classes that wish to follow this pattern then inherit from the abstract class and implement its methods – Java's interfaces provide similar functionality. Like in C, the linker builds applications by matching class and function names. Finally, virtual methods provide a more convenient and more structured way than function pointers for delaying beyond link-time decisions about what code to execute.

In nesC, programs are built out of a set of cooperating components. Each component uses interfaces to specify the services it provides and uses; the programmer uses wiring to build an application out of components by writing wiring statements, each of which connects an interface used by one component to an interface provided by another. Making these wiring statements explicit instead of relying on implicit name matching eliminates the requirement to use dynamic mechanisms (function pointers, virtual methods) to express concepts such as callbacks from a service to a client.

Part II

Basic programming

3 Components and interfaces

This chapter describes components, the building blocks of nesC programs. Every component has a signature, which describes the functions it needs to call as well as the functions that others can call on it. A component declares its signature with interfaces, which are sets of functions for a complete service or abstraction. Modules are components that implement and call functions in C-like code. Configurations connect components into larger abstractions. This chapter focuses on modules, and covers configurations only well enough to modify and extend existing applications: Chapter 4 covers writing new configurations from scratch.

3.1 Component signatures

A nesC program is a collection of components. Every component is in its own source file, and there is a one-to-one mapping between component and source file names. For example, the file LedsC.nc contains the nesC code for the component LedsC, while the component PowerupC can be found in the file PowerupC.nc. Components in nesC reside in a global namespace: there is only one PowerupC definition, and so the nesC compiler loads only one file named PowerupC.nc.

There are two kinds of components: modules and configurations. Modules and configurations can be used interchangeably when combining components into larger services or abstractions. The two types of components differ in their implementation sections. Module implementation sections consist of nesC code that looks like C. Module code declares variables and functions, calls functions, and compiles to assembly code. Configuration implementation sections consist of nesC *wiring* code, which connects components together. Configurations are the major difference between nesC and C (and other C derivatives).

All components have two code blocks. The first block describes its signature, and the second block describes its implementation:

```
module PowerupC {              configuration LedsC {
  // signature                   // signature
}                              }
implementation {               implementation {
  // implementation              // implementation
}                              }
```

Listing 3.1 The signature and implementation blocks

Signature blocks in modules and configurations have the same syntax. Component signatures contain zero or more interfaces. Interfaces define a set of related functions for a service or abstraction. For example, there is a Leds interface for controlling node LEDs, a Boot interface for being notified when a node has booted, and an Init interface for initializing a component's state. A component signature declares whether it **provides** or **uses** an interface. For example, a component that needs to turn a node's LEDs on and off uses the Leds interface, while the component that implements the functions that turns them on and off provides the Leds interface. Returning to the two examples, these are their signatures:

```
module PowerupC {                         configuration LedsC {
  uses interface Boot;                      provides interface Leds;
  uses interface Leds;                    }
}
```

Listing 3.2 Signatures of PowerupC and LedsC

PowerupC is a module that turns on a node LED when the system boots. As we saw in Chapter 2, it uses the Boot interface for notification of system boot and the Leds interface for turning on a LED. LedsC, meanwhile, is a configuration which provides the abstraction of three LEDs that can be controlled through the Leds interface. A single component can both provide and use interfaces. For example, this is the signature for the configuration MainC:

```
configuration MainC {
  provides interface Boot;
  uses interface Init;
}
```

Listing 3.3 MainC's signature

MainC is a configuration which implements the boot sequence of a node. It provides the Boot interface so other components, such as PowerupC, can be notified when a node has fully booted. MainC uses the Init interface so it can initialize software as needed before finishing the boot sequence. If PowerupC had a state that needed initialization before the system boots, it might provide the Init interface.

3.1.1 Visualizing components

Throughout this book, we'll use a visual language to show components and their relationships. Figure 3.1 shows the three components we've seen so far: MainC, PowerupC, and LedsC.

3.1 Component signatures 23

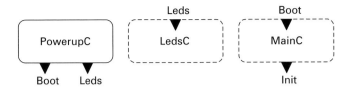

Figure 3.1 PowerupC, LedsC, and MainC. Triangles are interfaces. Triangles pointing out from a component are interfaces it uses, while triangles inside a component are interfaces it provides. A solid box is a module, while a dashed box is a configuration.

3.1.2 The "as" keyword and clustering interfaces

The **as** keyword lets a signature provide an alternative name for an interface. For example, MainC uses the **as** keyword to make its signature a bit clearer to the reader by using the name SoftwareInit for its Init interface:

```
uses interface Init as SoftwareInit;
```

Some signatures must use the keyword to distinguish multiple instances of the same interface. If a component provides or uses an interface more than once, its signature must use the **as** keyword to give them distinct names. For example, LedsC provides the abstraction of three LEDs through the Leds interface, but it is a configuration and not executable code. The LedsC configuration connects the LEDs module, LedsP, to components that provides the digital input-output lines which power the LEDs. The signature for LedsP is as follows:

```
module LedsP {
  provides {
    interface Init;
    interface Leds;
  }
  uses {
    interface GeneralIO as Led0;
    interface GeneralIO as Led1;
    interface GeneralIO as Led2;
  }
}
```

Listing 3.4 The LedsP module

A signature only needs to make sure that each interface instance has a unique name. For example, the LedsP example above could use **as** only twice, and leave one interface instance as GeneralIO, so the three would have the names Led0, Led1, and GeneralIO. However, in this case that would be confusing, so LedsP renames all three instances of GeneralIO. Technically, interface declarations have an implicit use of **as**. The statement

```
uses interface Leds;
```

is really shorthand for

```
uses interface Leds as Leds;
```

Generally, the keyword **as** is a useful tool for making components and their requirements clearer, similarly to how variable and function names greatly affect code readability.

Programming Hint 1 USE THE "AS" KEYWORD LIBERALLY.

3.1.3 Clustering interfaces

The LedsP example shows one further detail about signatures: they can cluster used and provided interfaces together. For example, these two versions of PowerupC are equivalent:

```
configuration PowerupC {          configuration PowerupC {
   uses interface Boot;              uses {
   uses interface Leds;                 interface Boot;
}                                       interface Leds;
                                     }
                                  }
```

Listing 3.5 PowerupC and an alternative signature

As these two are equivalent, there is no syntactical or code efficiency advantage to either approach: it is a matter of style and what is more legible to the reader. Often component signatures declare the interfaces they provide first, followed by the interfaces they use. This lets a reader clearly see the available functionality and dependencies. For very complex components that perform many functions, however, this approach breaks down, and signatures place related interfaces close to one another.

TinyOS detail: The names of all of the components described above end in the letters C and P. This is not a requirement. It is a coding convention used in TinyOS code. Components whose names end in C are abstractions that other components can use freely: the C stands for "component." Some component names end in P, which stands for "private." In TinyOS, P components should not be used directly, as they are generally an internal part of a complex system. Components use these two letters in order to clearly distinguish them from interfaces.

3.2 Interfaces

Interfaces describe a functional relationship between two or more different components. The role a component plays in this relationship depends on whether it provides or uses the interface. Like components, interfaces have a one-to-one mapping between names and files: the file Leds.nc contains the interface Leds while the file Boot.nc contains

3.2 Interfaces

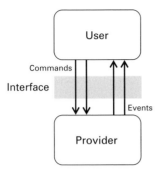

Figure 3.2 Interfaces have commands and events. Users call commands, and providers signal events.

the interface Boot. Just as with components, interfaces are in a global namespace. Syntactically, however, interfaces are quite different from components. They have a single block, the interface declaration:

```
interface Boot {              interface Leds {
  // functions                  // functions
}                             }
```

Listing 3.6 Interface declarations for Leds and Boot

An interface declaration has one or more functions in it. Interfaces have two kinds of functions: *commands* and *events*. Init and Boot are two simple interfaces, each of which has a single function. Init has a single command, while Boot has a single event:

```
interface Init {              interface Boot {
  command error_t init();       event void booted();
}                             }
```

Listing 3.7 The Init and Boot interfaces

TinyOS detail: The error_t type returned by init is TinyOS's normal way of reporting success or failure. A value of SUCCESS represents success and FAIL represents general failure. Specific Exxx constants, inspired in part by Unix's errno, represent specific failures, e.g. EINVAL means "invalid value".

Whether a function is a command or event determines which side of an interface – a user or a provider – implements the function and which side can call it. Users can **call** commands and providers can **signal** events. Conversely, users must implement events and providers must implement commands. Figure 3.3 shows this relationship in the visual language we use to describe nesC programs. For example, returning to MainC

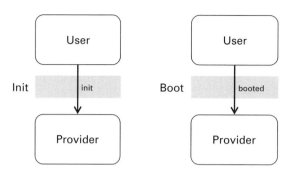

Figure 3.3 The Init and Boot interfaces.

and PowerupC, PowerupC is a user of Boot, while MainC is a provider of Boot:

```
configuration MainC {                      module PowerupC {
  provides interface Boot;                   uses interface Boot;
  uses interface Init as SoftwareInit;       uses interface Leds;
}                                          }
```

Listing 3.8 Signatures of MainC and PowerupC

Boot has a single event, Boot.booted. MainC provides Boot, so it signals the event when a node has booted successfully. This signaling is a function call: the boot sequence calls Boot.booted on all of the components that want to know when the system has come up. If a component – such as PowerupC – wants to know when the node has booted, it uses the Boot interface.

Commands have the opposite implementation direction as events. Users call commands and providers implement them. MainC uses Init so it can initialize software components before booting: it calls SoftwareInit.init on the components that need initialization. Similarly, as PowerupC uses Leds, it can call commands such as Leds.led0On. Conversely, a provider of Leds, such as LedsP, must implement the Leds commands.

Figure 3.3 shows the Boot and Init interfaces graphically. Init has a single command, init. A user of the Init interface can call Init.init, while a provider of the Init interface must implement Init.init. Conversely, a user of the Boot interface must implement the booted event, while a provider of the interface can signal Boot.booted.

At this point, the distinction between commands and events might seem pointless. After all, providing an interface with a command is the same as using an interface with an event. In the case of interfaces which only have commands or events but not both, this is true, and distinction is more for presentation and legibility. Providing an interface means providing a service, whether that be controlling the LEDs on a node or notifying components when the system has booted. Using an interface means using that service. However, as we'll soon see, many interesting interfaces have both commands and events, representing a relationship between two components where both can call the other.

3.2.1 Generic interfaces

The interfaces we've seen so far – Boot, Init, and Leds – are all type-free. The nesC language also supports *generic interfaces*, which take one or more types as a parameter. For example, the Queue interface takes a single parameter, which defines what it is a queue of:

```
interface Queue<t> {
  command bool empty();
  command uint8_t size();
  command uint8_t maxSize();
  command t head();
  command t dequeue();
  command error_t enqueue(t newVal);
  command t element(uint8_t idx);
}
```

Listing 3.9 The Queue interface

The Queue<t> statement says that the queue interface takes a single type, t. The interface definition can then refer to this type: enqueue takes a t as a parameter, while head, dequeue, and element return values of type t. Interfaces can have more than one type, in which case the types are comma-delimited: <a, b>.

When a component declares a generic interface, it must specify its parameters. For example, if a component needs to use a queue of 32-bit integers, its signature would include

```
module QueueUserC {
  uses interface Queue<uint32_t>;
}
```

Listing 3.10 Using a queue of 32-bit integers

Finally, when connecting users to providers, interface types must match. For example, QueueUserC can be connected to Queue32C below, but not Queue16C:

```
module Queue16C {                    module Queue32C {
  provides interface Queue<uint16_t>;  provides interface Queue<uint32_t>;
}                                    }
```

Listing 3.11 Providing a 16-bit or a 32-bit queue

Generic interfaces prevent unnecessary code duplication. Without generic interfaces, for example, we'd either need a separate Queue interface for every possible type needed, or Queue would have to take a generic type that a program casts to/from, such as a void*.

The former has the problem of code duplication (and file bloat), while the latter depends on run-time checking, which is notably deficient in C. Having a generic queue enables compile-time checking for the values put in and out of a queue.

3.2.2 Bidirectional interfaces

So far, we've only seen interfaces that have either commands or events, but not both. Bidirectional interfaces declare both commands from a user to a provider as well as events from a provider to a user. For example, this is the Notify interface, which allows a user to ask that it be notified of events, which can have data associated with them:

```
interface Notify<val_t> {
  command error_t enable();
  command error_t disable();
  event void notify(val_t val);
}
```

Listing 3.12. The Notify interface

The Notify interface has two commands, for enabling and disabling notifications. If notifications are enabled, then the provider of the interface signals notify events. The Notify interface is generic as, depending on the service, it might need to provide different kinds of data. Bidirectional interfaces enable components to register callbacks without needing function pointers.

For instance, some hardware platforms have a button on them. A button lends itself well to the Notify interface: a component can turn notifications of button pushes on and off. For example, UserButtonC is a component that provides this abstraction:

```
configuration UserButtonC {
  provides interface Get<button_state_t>;
  provides interface Notify<button_state_t>;
}
```

Listing 3.13 UserButtonC

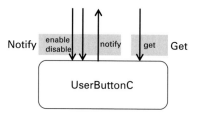

Figure 3.4 Commands and events for UserButtonC.

In addition to the Notify interface, which tells a user when the button state has changed, UserButtonC provides the Get interface, which a component can use to actively query the state of the button. Figure 3.4 shows the call directions of the commands and events of UserButtonC.

A component that provides Notify must implement the commands enable and disable, and can signal the notify event. A component that uses Notify can call the enable and disable commands, and must implement the notify event. In the case of UserButtonC, button_state_t describes whether the button is up or down.

Timer is another bidirectional interface. Timer is a generic interface, but it uses types in a slightly different way than Get or Notify. The type in a Timer interface is not used in any of its commands or events. Instead, the type represents a timer granularity. For example a Timer<TMilli> is a millisecond timer,[1] while a Timer<TMicro> is a microsecond timer. Typing in this way enables nesC to check that the Timer a component uses is the right granularity while only having a single Timer interface: you cannot wire a microsecond timer to a component that needs a millisecond timer. This is a subset of the Timer interface; it has additional, advanced operations which are elided for simplicity:

```
interface Timer<precision_tag> {
  command void startPeriodic(uint32_t dt);
  command void startOneShot(uint32_t dt);
  command void stop();
  event void fired();
  // Advanced operations follow
}
```

Listing 3.14 Simplified Timer interface showing three commands and one event

The fired event signals in response to the start commands, which differ in whether they cause a series of timer firings (startPeriodic) or a single fired event (startOneShot). The dt parameter specifies the timer interval.

Timer differs from Notify in a subtle but significant way: the user controls the timing and number of events. Notify allows users to turn events on and off; Timer allows users to control which events are signaled and when. There is therefore a much tighter coupling between the commands and events. With Notify, it's possible that no events will ever be signaled even if Notify is enabled. With Timer, one can describe exactly what events will be signaled based on what commands are called.

3.3 Component implementations

Modules and configurations differ in their implementation sections.

---

[1] Note that a TMilli timer fires 1024 times per second, not 1000. This is due to the fact that many microcontrollers do not have the ability to count at 1 kHz accurately, but they can count at 1024 Hz accurately.

3.3.1 Modules

Module implementations are written in nesC's dialect of the C language. At the level of single statements, nesC is almost identical to C. It differs mainly in how code calls functions and variable scoping. As a component must implement commands for interfaces it provides and events for interfaces it uses, a module's code must define these functions. For example, this is the module code for the PowerupC module:

```
module PowerupC {
  uses interface Boot;
  uses interface Leds;
}
implementation {
  event void Boot.booted() {
    call Leds.led0On();
  }
}
```

Listing 3.15 PowerupC module code

As PowerupC uses the Boot interface, and Boot has a single event, booted, PowerupC must provide an implementation: `event void Boot.booted() { ... }`. When the provider of Boot signals this event, PowerupC turns on LED 0 of the Leds interface. Modules invoke commands with the **call** keyword and invoke events with the **signal** keyword. For this reason, you'll often see the terms "call commands" and "signal events."

The Leds interface has three LEDs, mostly for historical reasons. They are named led0, led1, and led2. In addition to turning LEDs on (e.g., led0On), the Leds interface allows you to turn them off (e.g., led0Off) and toggle them (e.g., led0Toggle). For example, this component will cause system boot to enter an infinite loop that toggles all three LEDs:

```
module PowerupToggleC {
  uses interface Boot;
  uses interface Leds;
}
implementation {
  event void Boot.booted() {
    while (1) {
      call Leds.led0Toggle();
      call Leds.led1Toggle();
      call Leds.led2Toggle();
    }
  }
}
```

Listing 3.16 The module PowerupToggleC

For PowerupToggleC to do anything, we must specify which components providing Boot and Leds we want to connect PowerupToggleC to. This is the role of the PowerupToggleAppC configuration which we will write next.

3.3.2 A basic configuration

Modules allocate state and implement executable logic. However, like all components, they can only name functions and variables within their local namespaces, i.e., those defined in their signature and implementation. For one module to be able to call another, we have to map a set of names (generally interfaces) in one component's signature to a set of names in another component signature. In nesC, connecting two components in this way is called wiring. Configurations, nesC's second kind of component, contain these wirings. Modules implement program logic, configurations connect components together into larger abstractions.

Configurations look very similar to modules. They have a specification and an implementation. This is the PowerupToggleAppC configuration, which connects the PowerupToggleC module to the components that provide the Boot and Leds interfaces.

```
configuration PowerupToggleAppC {}
implementation {
  components MainC, LedsC, PowerupToggleC;

  PowerupToggleC.Boot -> MainC.Boot;
  PowerupToggleC.Leds -> LedsC.Leds;
}
```

Listing 3.17 The PowerupToggleAppC configuration

A configuration must name which components it is wiring with the **components** keyword. Any number of component names can follow **components**, and their order does not matter. The keyword remains plural (**components**, not component) even if only a single component name follows. A configuration can have multiple **components** statements. A configuration must name a component before it wires it. For example, both of these are valid:

```
components MainC, LedsC;
components PowerupToggleC;
PowerupToggleC.Boot -> MainC.Boot;
PowerupToggleC.Leds -> LedsC.Leds;

components PowerupToggleC;
components MainC;
PowerupToggleC.Boot -> MainC.Boot;
components LedsC;
PowerupToggleC.Leds -> LedsC.Leds;
```

Listing 3.18 Example uses of the components keyword

Syntactically, configurations are very simple. They have three operators: ->, <- and =. The = operator is used for wiring the configuration's specification, as we will see in

Chapter 4. The two arrows are for wiring a configuration's components to each other: the arrow connects an interface user to an interface provider. The arrow points from the user to the provider, but resolves the call paths of bidirectional interfaces in both directions (used-to-provided commands, and provided-to-used events). For example, the following two lines have the same effect:

```
PowerupToggleC.Boot -> MainC.Boot;
MainC.Boot <- PowerupToggleC.Boot;
```

When PowerupToggleC calls Leds.led0Toggle, it names a function in its own local scope. The LedsC component provides the Leds interface. Wiring the two, maps the first to the second. This means that when PowerupToggleC calls its Leds.led0Toggle, it actually calls LedsC's Leds.led0Toggle. The same is true for other calls of the Leds interface, such as Leds.led1On. The configuration PowerupToggleAppC provides a mapping between the local namespaces of the two components.

Because PowerupToggleAppC is the top-level configuration of the PowerupToggle application, it does not provide or use any interfaces: its signature block is empty. Later, in Chapter 4, we'll introduce configurations that provide and use interfaces.

3.3.3 Module variables

All module variables are private: interfaces are the only way that other components can access a variable. The Get interface, mentioned above as part of the UserButtonC, is an example of such an abstraction. Get has a very simple definition:

```
interface Get<val_t> {
  command val_t get();
}
```

Listing 3.19 The Get interface

Modules declare variables much like standard C. For example, this component implements a Get interface which returns the number of times get has been called (i.e. acts like a counter):

```
module CountingGetC {
  provides interface Get<uint8_t>;
}
implementation {
  uint8_t count;

  command uint8_t Get.get() {
    return count++;
  }
}
```

Listing 3.20 A self-incrementing counter

Module variable declarations can have initializers, just like C:

```
uint8_t count = 1;
message_t packet;
message_t* packetPtr = &packet;
```

3.3.4 Generic components

By default, components in TinyOS are *singletons*: only one exists. Every configuration that names a singleton component names the same component. For example, if two configurations wire to LedsC, they are wiring to the same code that accesses the same variables. A singleton component introduces a component name that any configuration can use into the global namespace.

In addition to singleton components, nesC has *generic components*. Unlike singletons, a generic component can have multiple instances. For example, while a low-level software abstraction of a hardware resource is inherently a singleton – there is only one copy of a hardware register – software data structures are instantiable. Being instantiable makes them reusable across many different parts of an application. For example, the module BitVectorC provides the abstraction of a bit vector; rather than define macros or functions to manipulate a bit vector a module can just use the interface BitVector and assume that a corresponding configuration connects it to a BitVectorC of the proper width.

Earlier versions of nesC (1.0 and 1.1) did not support generic components. Whenever a component requires a common data structure, a programmer had to make a copy of the data structure component and give it a new name, or separate functionality and allocation by locally allocating data structures and using library routines. For example, network protocols typically all implemented their own queue data structures, rather than relying on a standard implementation. This code copying prevented code reuse, forcing programmers to continually revisit common bugs and problems, rather than building on well-tested libraries.

Generic components have the keyword **generic** before their signature:

```
generic module SineSensorC() {          generic configuration TimerMilliC() {
  provides interface Init;                provides interface Timer<TMilli>;
  provides interface Read<uint16_t>;    }
}
```

Listing 3.21 Generic module SineSensorC and generic configuration TimerMilliC

To use a generic component, a configuration must instantiate it with the **new** keyword. This is the beginning of the code for the configuration BlinkAppC, the top-level configuration for the Blink application, which displays a 3-bit counter on a mote's LEDs using three timers:

```
configuration BlinkAppC {}
implementation {
  components MainC, BlinkC, LedsC;
  components new TimerMilliC() as Timer0;
  components new TimerMilliC() as Timer1;
  components new TimerMilliC() as Timer2;
  /* Wirings below */
}
```

Listing 3.22 Instantiating a generic component

Generic components can take parameters, hence the parentheses in component signatures (`generic configuration TimerMilliC()`) and instantiations (`components new TimerMilliC() as Timer0;`). These parameters can be values of simple types, constant strings, or types. For example, BitVectorC takes a 16-bit integer denoting how many bits there are:

```
generic module BitVectorC(uint16_t maxBits) {
  provides interface Init;
  provides interface BitVector;
}
```

Listing 3.23 Signature of BitVectorC

The **typedef** keyword denotes a parameter to a generic component that is a type. The generic module QueueC is a queue with a fixed maximum length. QueueC takes two parameters: the type that the queue stores and the maximum length. By convention, we suffix all type arguments with _t:

```
generic module QueueC(typedef queue_t, uint8_t queueSize) {
  provides interface Queue<queue_t>;
}
```

Listing 3.24 QueueC signature

Chapter 8 goes into the details of writing new generic components.

3.4 Split-phase interfaces

Because sensor nodes have a broad range of hardware capabilities, one of the goals of TinyOS is to have a flexible hardware/software boundary. An application that encrypts packets should be able to interchangeably use hardware or software implementations. Hardware, however, is almost always *split-phase* rather than blocking. In a *split-phase* operation the request that initiates an operation completes immediately.

3.4 Split-phase interfaces

Actual completion of the operation is signaled by a separate callback. For example, to acquire a sensor reading with an analog-to-digital converter (ADC), software writes to a few configuration registers to start a sample. When the ADC sample completes, the hardware issues an interrupt, and the software reads the value out of a data register.

Now, let's say that rather than directly sampling, the sensor implementation actually samples periodically and when queried gives a cached value. This may be necessary if the sensor needs to continually calibrate itself. Magnetometer drivers sometimes do this due to the effect of the Earth's magnetic field, as two sensors oriented differently might have very different magnetometer floors. Drivers estimate the floor and essentially return a measure of recent change, rather than an absolute value. From a querying standpoint, the implementation of the sensor is entirely in software. This fact should not be apparent to the caller. For ease of composition, sampling a self-calibrating magnetometer should be the same as a simple photoresistor. But the magnetometer is a synchronous operation (it can return the result immediately) while the ADC is split-phase.

The basic solution to this problem is to make one of the two look like the other: either give the magnetometer a split-phase interface, or make the ADC synchronous by blocking until the sampling completes. If the ADC interrupt is very fast, the ADC driver might be able to get away with a simple spin loop to wait until it fires. If the interrupt is slow, then this wastes a lot of CPU cycles and energy. The traditional solution for this latter case (e.g. in traditional operating systems) is to use multiple threads. When the code requests an ADC sample, the OS sets up the request, puts the calling thread on a wait queue, starts the operation, and then schedules another thread to run. When the interrupt comes in, the driver resumes the waiting thread and puts it on the OS ready queue.

The problem with threads in embedded systems is that they require a good deal of RAM. Each thread has its own private stack which has to be stored when a thread is waiting or idle. For example, when a thread samples a blocking ADC and is put on the wait queue, the memory of its entire call stack has to remain untouched so that when it resumes it can continue execution. RAM is a very tight resource on current sensor node platforms. Early versions of TinyOS ran in 512 bytes of RAM. When a thread is idle, its stack is wasted storage, and allocating the right sized stack for all of the threads in the system can be a tricky business. Additionally, while it is easy to layer threads on top of a split-phase interface, it is very difficult to do the opposite. Because it's a one-way street, while increasing amounts of RAM might allow threads at an application level, the bottom levels of TinyOS – the core operating system – can't require them, as they preclude chips with leaner RAM resources than high-end micro controllers.

TinyOS therefore takes the opposite approach. Rather than make everything synchronous through threads, operations that are split-phase in hardware are split-phase in software as well. This means that many common operations, such as sampling sensors and sending packets, are split-phase. An important characteristic of split-phase interfaces is that they are bidirectional: there is a command to start the operation, and an event that signifies the operation is complete. As a result, TinyOS programs only need a single stack, saving RAM.

3.4.1 Read

The Read interface is the basic TinyOS interface for split-phase data acquisition. Most sensor drivers provide Read, which is generic:

```
interface Read<val_t> {
  command error_t read();
  event void readDone(error_t err, val_t val);
}
```

Listing 3.25 The Read interface

For example, a sensor driver that generates a 16-bit value provides Read<uint16_t>. If the provider of Read returns SUCCESS to a call to read, then it will signal readDone in the future, passing the Read's result back as the val parameter to the event handler.

3.4.2 Send

The basic TinyOS packet transmission interface, Send, is also a split-phase operation. However, it is slightly more complex, as it requires passing a pointer for a packet to transmit:

```
interface Send {
  command error_t send(message_t* msg, uint8_t len);
  event void sendDone(message_t* msg, error_t error);

  command error_t cancel(message_t* msg);
  command void* getPayload(message_t* msg);
  command uint8_t maxPayloadLength(message_t* msg);
}
```

Listing 3.26 The split-phase Send interface

A provider of Send defines the send and cancel functions and can *signal* the sendDone event. Conversely, a user of Send needs to define the sendDone event and can *call* the send and cancel commands. When a call to send returns SUCCESS, the msg parameter has been passed to the provider, which will try to send the packet. When the send completes, the provider signals sendDone, passing the pointer back to the user. This pointer passing approach is common in split-phase interfaces that need to pass larger data items. The next section discusses memory management in greater detail.

3.5 Module memory allocation, avoiding recursion, and other details

Besides power, the most valuable resource to mote systems is RAM. Power means that the radio and CPU have to be off almost all the time. Of course, there are situations which need a lot of CPU or a lot of bandwidth (e.g. cryptography or binary dissemination), but

3.5 Memory allocation, avoiding recursion

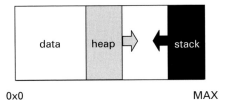

Figure 3.5 Typical memory layout on a microcontroller. Because there is no memory protection, the stack can easily overflow onto the heap or data.

by necessity they have to be rare occurrences. In contrast, the entire point of RAM is that it's always there. The sleep current of the microcontrollers most motes use today is, for the most part, determined by RAM.

Modules allocate memory by declaring variables which, following nesC's scoping rules, are completely private to the component. For example, the CountingGetC component (Listing 3.20, page 32) allocated count as an 8-bit module-level variable, for a cost of 1 byte of RAM. Because TinyOS uses split-phase operations and does not provide threads, there is no long-lived stack-allocated data. As a result, when a TinyOS system is quiescent, these module variables represent the entire software state of the system.

Generally, nesC does not encourage dynamic memory allocation through malloc or other C library calls. You can call them, but the lack of memory protection on most embedded microcontrollers makes their use particularly risky. Figure 3.5 shows a typical memory layout on a microcontroller. The stack grows down and the heap grows up, and since there is no hardware memory protection the two can collide, at which point chaos is guaranteed.

Instead, you should allocate memory as module variables. For example, if a module needs a buffer with which to hold sensor readings, it should allocate the buffer statically. In other cases, it is convenient to create reusable abstract data types by packaging up some state and operations in a generic component, as in BitVectorC (Listing 3.23, page 34) . Finally, components sometimes need to share a memory pool. A common example of this is a set of components that share a pool of packet buffers. A shared pool allows multiple cooperating components to amortize their requirements, especially if it is unlikely all of them will need a lot of memory at the same time. By avoiding all use of the heap, the only cause of run-time memory failure is the stack.

To avoid stack overflow, TinyOS programs should avoid recursion and not declare any large local variables (e.g. arrays). Avoiding recursion within a single module is easy, but in a component-based language like nesC it's very easy to unintentionally create a recursive loop across component boundaries. For instance, let's assume component A uses the Read interface to repeatedly sample a sensor provided by component B, i.e. the readDone event handler in A calls B's read command. If B happens to be a simple sensor, it might choose to signal the readDone event directly within the implementation of the read command. However, this program now contains an unintended recursive loop: A calls B's read command which signals A's readDone event which calls B's read command which ...

To avoid such recursive loops, TinyOS follows a couple of coding conventions. First, split-phase commands must never directly signal their callback – see Section 5.3.2 for more details. Second, the relation between most TinyOS components is hierarchical: application components use interfaces provided by system services, which themselves use interfaces provided by lower-level services, and so on down to the raw hardware – this structure is discussed in depth in Chapter 12.

Finally, it's worth noting that the stack may overflow because of extra stack usage caused by an interrupt handler (Chapter 5) interrupting your regular computation (or, even worse, another interrupt handler which is already using some of the stack space). You should always leave enough RAM free to handle the worst case usage of your regular computation and all interrupt handlers that can execute simultaneously.

> **Programming Hint 2** NEVER WRITE RECURSIVE FUNCTIONS WITHIN A MODULE. IN COMBINATION WITH THE TINYOS CODING CONVENTIONS, THIS GUARANTEES THAT ALL PROGRAMS HAVE BOUNDED STACK USAGE.

> **Programming Hint 3** NEVER USE MALLOC AND FREE. ALLOCATE ALL STATE IN COMPONENTS. IF YOUR APPLICATION REQUIREMENTS NECESSITATE A DYNAMIC MEMORY POOL, ENCAPSULATE IT IN A COMPONENT AND TRY TO LIMIT THE SET OF USERS.

3.5.1 Memory ownership and split-phase calls

TinyOS programs contain many concurrent activities, e.g. even in a very simple program, radio message transmission, sensor sampling and application logic. Ensuring that these activities do not step on each other by accessing each other's data out of turn is often a complex problem.

The only way that components can interact is through function calls, which are normally part of interfaces. Just as in C, there are two basic ways that components can pass parameters: by value and by reference (pointer). In the first case, the data is copied onto the stack, so the callee can modify it or cache it freely. In the second case, the caller and callee share a pointer to the data, and so the two components need to carefully manage access to the data in order to prevent memory corruption.

The simplest solution to preventing data-sharing problems is to never store pointer parameters in module variables. This is the approach used by some abstract data type components (see Section 8.2.2); it ensures that any data-sharing is transitory, restricted to the duration of the command or event with the pointer parameter.

However, this approach is not practical for split-phase calls. Because the called component typically needs access to the pointer while the operation is executing, it has to store it in a module variable. For example, consider the basic Send interface:

```
interface Send {
  command error_t send(message_t* msg, uint8_t len);
  event void sendDone(message_t* msg, error_t error);
```

3.5 Memory allocation, avoiding recursion

```
  command error_t cancel(message_t* msg);
  command void* getPayload(message_t* msg);
  command uint8_t maxPayloadLength(message_t* msg);
}
```

Listing 3.27 The Send interface

The important pair of functions in this example is send/sendDone. To send a packet, a component calls send. If send returns SUCCESS, then the caller has passed the packet to a communication stack to use, and must not modify the packet. The callee stores the pointer in a variable, enacts a state change, and returns immediately. If the interface user modifies the packet after passing it to the interface provider, the packet could be corrupted. For example, the radio stack might compute a checksum over the entire packet, then start sending it out. If the caller modifies the packet after the checksum has been calculated, then the data and checksum won't match up and a receiver will reject the packet.

To avoid these kinds of problems, TinyOS follows an *ownership* discipline: at any point in time, every "memory object" – a piece of memory, typically a whole variable or a single array element – should be owned by a single module. A command like send is said to *pass ownership* of its msg argument from caller to callee. When a split-phase interface has this kind of "pass" semantics, the completion event should have the passed pointer as one of its parameters, to show that the object is being returned to its original owner.

Programming Hint 4 WHEN POSSIBLE, AVOID PASSING POINTERS ACROSS INTERFACES; WHEN THIS CANNOT BE AVOIDED ONLY ONE COMPONENT SHOULD BE ABLE TO MODIFY A POINTER'S DATA AT ANY TIME.

One of the trickiest examples of this pass approach is the Receive interface. At first glance, the interface seems very simple:

```
interface Receive {
  event message_t* receive(message_t* msg, void* payload, uint8_t len);
}
```

Listing 3.28 The Receive interface

The receive event is rather different than most events: it has a message_t* as both a parameter and a return value. When the communication layer receives a packet, it passes that packet to the higher layer as a parameter. However, it also expects the higher layer to return it a message_t* back. The basic idea behind this is simple: if the communication layer doesn't have a message_t*, it can't receive packets, as it has nowhere to put them. Therefore, the higher layer always has to return a message_t*, which is the next buffer the radio stack will use to receive into. This return value can be the same as the parameter, but it does not have to be. For example, this is perfectly reasonable, if a bit feature-free, code:

```
event message_t* Receive.receive(message_t* msg, void* payload, uint8_t len) {
   return msg;
}
```

A receive handler can always copy needed data out of the packet and just return the passed buffer. There are, however, situations when this is undesirable. One common example is a routing queue. If the node has to forward the packet it just received, then copying it into another buffer is wasteful. Instead, a queue allocates a bunch of packets, and in addition to a send queue, keeps a free list. When the routing layer receives a packet to forward, it sees if there are any packets left in the free list. If so, it puts the received packet into the send queue and returns a packet from the free list, giving the radio stack a buffer to receive the next packet into. If there are no packets left in the free list, then the queue can't accept the packet and so just returns it back to the radio for re-use. The pseudocode looks something like this:

```
receive (m):
    if I'm not the next hop, return m    // Not for me
    if my free list is empty, return m   // No space
    else
        put m on forwarding queue
        return entry from free list
```

One of the most common mistakes early TinyOS programmers encounter is misusing the Receive interface. For example, imagine a protocol that does this:

```
event message_t* LowerReceive.receive(message_t* m, void* payload, uint8_t len) {
   processPacket(m);
   if (amDestination(m)) {
      signal UpperReceive.receive(m, payload, len);
   }
   return m;
}
```

The problem with this code is that it ignores the return value from the signal to UpperReceive.receive. If the component that handles this event performs a buffer swap – e.g. it has a forwarding queue – then the packet it returns is lost. Furthermore, the packet that it has put on the queue has also been returned to the radio for the next packet reception. This means that, when the packet reaches the end of the queue, the node may send something completely different than what it decided to forward (e.g. a packet for a completely different protocol).

The buffer swap approach of the Receive interface provides isolation between different communication components. Imagine, for example, a more traditional approach, where the radio dynamically allocates a packet buffer when it needs one. It allocates buffers and passes them to components on packet reception. What happens if a component holds on to its buffers for a very long time? Ultimately, the radio stack will run out of memory to allocate from, and will cease being able to receive packets at all. By pushing the allocation policy up into the communication components, protocols that have no free memory left are forced to drop packets, while other protocols continue unaffected.

This approach speaks more generally of how nesC components generally handle memory allocation. All state is allocated in one of two places: components, or the stack. A shared dynamic memory pool across components makes it much easier for one bad component to cause others to fail. That is not to say that dynamic allocation is never used. For example, the PoolC component provides a memory pool of a fixed number of a single type. Different components can share a pool, dynamically allocating and deallocating as needed:

```
generic configuration PoolC(typedef pool_t, uint8_t POOL_SIZE) {
  provides interface Pool<pool_t>;
}
```

Listing 3.29 The signature of PoolC

Bugs or resource exhaustion in components using a particular pool do not affect components using a different, or no, pool.

3.5.2 Constants and saving memory

Modules often need constants of one kind or another, such as a retransmit count or a threshold. Using a literal constant is problematic, as you'd like to be able to reuse a consistent value. This means that in C-like languages, you generally use something like this:

```
const int MAX_RETRANSMIT = 5;

if (txCount < MAX_RETRANSMIT) {
  ...
}
```

The problem with doing this in nesC/TinyOS is that a const int might allocate RAM, depending on the compiler (good compilers will place it in program memory). You can get the exact same effect by defining an enum:

```
enum {
  MAX_RETRANSMIT = 5
};
```

This allows the component to use a name to maintain a consistent value and does not store the value either in RAM or program memory. This can even improve performance, as rather than a memory load, the architecture can just load a constant. It's also better than a #define, as it exists in the debugging symbol table and application metadata. However, enum can only declare integer constants, so you should still use #define for floating-point and string constants (but see Section 3.5.5 for a discussion of some of #define's pitfalls).

Note, however, that using enum types in variable declarations can waste memory, as enums default to integer width. For example, imagine this enum:

```
typedef enum {
  STATE_OFF = 0,
  STATE_STARTING = 1,
  STATE_ON = 2,
  STATE_STOPPING = 3
} state_t;
```

Here are two different ways you might allocate the state variable in question:

```
state_t state; // platform int size (e.g., 2-4 bytes)
uint8_t state; // one byte
```

Even though the valid range of values is 0–3, the former will allocate a native integer, which on a microcontroller is usually 2 bytes, but could be 4 bytes on low-power microprocessors. The second will allocate a single byte. So you should use enums to declare constants, but avoid declaring variables of an enum type.

> **Programming Hint 5** CONSERVE MEMORY BY USING ENUMS RATHER THAN CONST VARIABLES FOR INTEGER CONSTANTS, AND DON'T DECLARE VARIABLES WITH AN ENUM TYPE.

3.5.3 Platform-independent types

To simplify networking code, TinyOS has traditionally used structs to define message formats and directly access messages – this avoids the programming complexity and overheads of using marshalling and unmarshalling functions to convert between host and network message representations. For example, the standard header of a packet for the CC2420 802.15.4 wireless radio chip[2] looks something like this:

```
typedef struct cc2420_header_t {
  uint8_t length;
  uint16_t fcf;
  uint8_t dsn;
  uint16_t destpan;
  uint16_t dest;
  uint16_t src;
  uint8_t type;
} cc2420_header_t;
```

Listing 3.30 CC2420 packet header

That is, it has a 1-byte length field, a 2-byte frame control field, a 1-byte sequence number, a 2-byte group, a 2-byte destination, a 2-byte source, and 1-byte type fields. Defining this as a structure allows you to easily access the fields, allocate storage, etc.

---

[2] Standard in that IEEE 802.15.4 has several options, such as 0-byte, 2-byte, or 8-byte addressing, and so this is just the format TinyOS uses by default.

3.5 Memory allocation, avoiding recursion

The problem, though, is that the layout and encoding of this structure depends on the chip you're compiling for. For example, the CC2420 expects all of these fields to be little-endian. If your microcontroller is big-endian, then you won't be able to easily access the bits of the frame control field. One commonly used solution to this problem is to explicitly call macros that convert between the microcontroller and the chip's byte order, e.g. macros like Unix's htons, ntohl, etc. However, this approach is error-prone, especially when code is initially developed on a processor with the same byte order as the chip.

Another problem with this approach is due to differing alignment rules between processors. On an ATmega128, the structure fields will be aligned on 1-byte boundaries, so the layout will work fine. On an MSP430, however, 2-byte values have to be aligned on 2-byte boundaries: you can't load an unaligned word. So the MSP430 compiler will introduce a byte of padding after the length field, making the structure incompatible with the CC2420 and other platforms. There are a couple of other issues that arise, but the eventual point is the same: TinyOS programs need to be able to specify platform-independent data formats that can be easily accessed and used.

In TinyOS 1.x, some programs attempted to solve this problem by using gcc's packed attribute to make data structures platform independent. Packed tells gcc to ignore normal platform struct alignment requirements and instead pack a structure tightly:

```
typedef struct RPEstEntry {
    uint16_t id;
    uint8_t receiveEst;
} _attribute_ ((packed)) RPEstEntry;
```

Listing 3.31 The dreaded "packed" attribute in the 1.x MintRoute library

Packed allowed code running on an ATmega128 and on an x86 to agree on data formats. However, packed has several problems. The version of gcc for the MSP430 family (used in Telos motes) doesn't handle packed structures correctly. Furthermore, packed is a gcc-specific feature, so code that uses it is not very portable. And finally, while packed eliminates alignment differences, it does not change endianness: int16_t maybe be big-endian on one platform and little-endian on another, so you would still have to use conversion macros like htons.

Programming Hint 6 NEVER, EVER USE THE "PACKED" ATTRIBUTE IN PORTABLE CODE.

To keep the convenience of specifying packet layouts using C types while keeping code portable, nesC 1.2 introduced platform-independent types. Simple platform-independent types (integers) are either big-endian or little-endian, independently of the underlying chip hardware. Generally, an external type is the same as a normal type except that it has nx_ or nxle_ preceding it:

```
nx_uint16_t val;    // A big-endian 16-bit value
nxle_uint32_t otherVal; // A little-endian 32-bit value
```

In addition to simple types, there are also platform-independent structs and unions, declared with **nx_struct** and **nx_union**. Every field of a platform-independent struct or union must be a platform-independent type. Non-bitfields are aligned on byte boundaries (bitfields are packed together on bit boundaries, as usual). For example, this is how TinyOS 2.0 declares the CC2420 header:

```
typedef nx_struct cc2420_header_t {
  nxle_uint8_t length;
  nxle_uint16_t fcf;
  nxle_uint8_t dsn;
  nxle_uint16_t destpan;
  nxle_uint16_t dest;
  nxle_uint16_t src;
  nxle_uint8_t type;
} cc2420_header_t;
```

Listing 3.32 The CC2420 header

Any hardware architecture that compiles this structure uses the same memory layout and the same endianness for all of the fields. This enables platform code to pack and unpack structures, without resorting to macros or utility functions such as UNIX socket htonl and ntohs.

Programming Hint 7 USE PLATFORM-INDEPENDENT TYPES WHEN DEFINING MESSAGE STRUCTURES.

Under the covers, nesC translates network types into byte arrays, which it packs and unpacks on each access. For most nesC codes, this has a negligible run-time cost. For example, this code

```
nx_uint16_t x = 5;
uint16_t y = x;
```

rearranges the bytes of x into a native chip layout for y, taking a few cycles. This means that if you need to perform significant computation on arrays of multibyte values (e.g. encryption), then you should copy them to a native format before doing so, then move them back to a platform-independent format when done. A single access costs a few cycles, but thousands of accesses costs a few thousand cycles.

Programming Hint 8 IF YOU HAVE TO PERFORM SIGNIFICANT COMPUTATION ON A PLATFORM-INDEPENDENT TYPE OR ACCESS IT MANY (HUNDREDS OR MORE) TIMES, TEMPORARILY COPY IT TO A NATIVE TYPE.

3.5.4 Global names

Components encapsulate functions and state, and wiring connects functions defined in different components. However, nesC programs also need globally available types for common abstractions, such as error_t (TinyOS's error code abstraction) or message_t

3.5 Memory allocation, avoiding recursion

(networking buffers). Furthermore, nesC programs sometimes call existing C library functions, either from the standard C library (e.g. mathematical functions like sin) or functions from a personal library of existing C code (see Section 3.5.6).

In keeping with C, nesC uses `.h` header files and `#include` for this purpose. This has the added advantage that existing C header files can be directly reused. For instance, TinyOS's error_t type and error constants are defined in the `TinyError.h` file:

```
#ifndef TINY_ERROR_H_INCLUDED
#define TINY_ERROR_H_INCLUDED

enum {
  SUCCESS     = 0,
  FAIL        = 1,   // Generic condition: backwards compatible
  ESIZE       = 2,   // Parameter passed in was too big.
  ECANCEL     = 3,   // Operation cancelled by a call.
  EOFF        = 4,   // Subsystem is not active
  EBUSY       = 5,   // The underlying system is busy; retry later
  EINVAL      = 6,   // An invalid parameter was passed
  ERETRY      = 7,   // A rare and transient failure: can retry
  ERESERVE    = 8,   // Reservation required before usage
  EALREADY    = 9,   // The device state you are requesting is already set
};

typedef uint8_t error_t;
#endif
```

Listing 3.33 TinyError.h, a typical nesC header file

Like a typical C header file, `TinyError.h` uses #ifndef/#define to avoid redeclaring the error constants and error_t when the file is included multiple times. Including a header file in a component is straightforward:

```
#include "TinyError.h"

module BehaviorC { ... }
implementation
{
  error_t ok = FAIL;
}
```

Listing 3.34 Including a header file in a component

Just as in C, #include just performs textual file inclusion. As a result it *is* important to use #include in the right place, i.e. before the **interface**, **module**, or **configuration** keyword. If you don't, you won't get the behavior you expect. Similarly, in C, using `#include <stdio.h>` in the middle of a function is not likely to work.

Unlike C where each file is compiled separately, constants, types, and functions included in one component or interface are visible in subsequently compiled components or interfaces. For instance, `TinyError.h` is included by interface Init, so the following module can use error_t, SUCCESS, etc:

```
module BadBehaviorC {
  provides interface Init;
}
implementation
{
  command error_t Init.init() {
    return FAIL; // We're bad, we always fail.
  }
}
```

Listing 3.35 Indirectly including a header file

Programming Hint 9 INTERFACES SHOULD #INCLUDE THE HEADER FILES FOR THE TYPES THEY USE.

Header files written for nesC occasionally include C function definitions, not just declarations. This is practical because the header file ends up being included exactly once in the whole program, unlike in C where it is included once per file (leading to multiple definitions of the same function). These uses are, however, rare, as they go against the goal of encapsulating all functionality within components.

3.5.5 nesC and the C preprocessor

Preprocessor symbols #defined before a nesC's file **module**, **configuration**, or **interface** keyword are available in subsequently loaded files, while those #defined later are "forgotten" at the end of the file:

```
// Available in all subsequently loaded files
#define GLOBAL_NAME "fancy"

interface Fancy {
// Forgotten at the end of this file
#define LOCAL_NAME "soon_forgotten"
  command void fancyCommand();
}
```

Listing 3.36 Fancy.nc: C preprocessor example

However, relying directly on this behavior is tricky, because the preprocessor is run (by definition) before a file is processed. Consider a module that uses the Fancy interface:

```
module FancyModule {
  uses interface Fancy;
}
implementation {
  char *name = GLOBAL_NAME;
}
```

Listing 3.37 FancyModule.nc: C preprocessor pitfalls

Compiling FancyModule will report that GLOBAL_NAME is an unknown symbol. Why? The problem is that the first step in compiling `FancyModule.nc` is to preprocess it. At that point, the Fancy interface hasn't been seen yet, therefore it hasn't been loaded and the GLOBAL_NAME #define is unknown. Later on, when FancyModule is analyzed, the Fancy interface is seen, the `Fancy.nc` file is loaded and GLOBAL_NAME gets #defined. But this is too late to use it in `FancyModule.nc`.

There are two lessons to be drawn from this: first, as we've already seen, it's best to use **enum** rather than #define to define constants when possible. Second, if you must use #define, use it as you would in C: place your definitions in a header file protected with #ifndef/#define, and #include this header file in all components and interfaces that use the #define symbol. For instance, both `Fancy.nc` and `FancyModule.nc` should

```
#include "Fancy.h"
```

where `Fancy.h` contains:

```
#ifndef FANCY_H
#define GLOBAL_NAME "fancy"
#endif
```

Listing 3.38 Fancy.h: the reliable way to use C preprocessor symbols

Programming Hint 10 ALWAYS #DEFINE A PREPROCESSOR SYMBOL IN A HEADER FILE. USE #INCLUDE TO LOAD THE HEADER FILE IN ALL COMPONENTS AND INTERFACES THAT USE THE SYMBOL.

3.5.6 C libraries

Accessing a C library is easy. The implementation of the generic SineSensorC component uses the C library's sin function, which is defined in the `math.h` header file:

```
#include <math.h>
generic module SineSensorC() {
  provides interface Init;
  provides interface Read<uint16_t>;
}
```

```
implementation {
  uint32_t counter;

  void performRead() {
    float val = sin(counter++ / 20.0);
    signal Read.readDone(SUCCESS, val * 32768.0 + 32768.0);
  }
  ...
}
```

Listing 3.39 Using a C library function

As with C, if you use a library, you also need to link with it. In this case, nesC programs using SineSensorC need to link with the math library by passing the -lm option to ncc, the nesC compiler driver.

3.6 Exercises

1. The Send interface has a command **send** and an event **sendDone**. Should you be able to call send in the sendDone handler or should you have to wait until after the handler returns? Why? Write pseudocode for the provider of both versions, and write pseudocode for a user of each version that wants to send packets as quickly as possible.
2. Extend the Blink application so it displays the bottom bits of a random number rather than a counter. Generate the random number from one of the random number generators in the TinyOS component libraries (RandomMlcgC or RandomLfsrC).
3. Write an application that increments a 32-bit counter in an infinite loop. Every N increments, the application toggles LED 0. Choose an N so you can observe the toggling visually. Try making the counter a platform-independent type. Does the toggling slow down? How much? Try a 16-bit value.
4. Write a PingPong application that runs on two nodes. When a node boots, it sends a broadcast packet using the AMSend interface. When it receives a packet, it sends a packet. Toggle an LED whenever a node sends a packet. How long does it take for the LED to stop toggling? Try different distances. If it's too fast to see, start a timer when the node receives a packet, and send a reply when the timer fires. You'll want to use the AMSenderC and AMReceiverC components.

4 Configurations and wiring

The previous chapter dealt predominantly with modules, which are the basic building blocks of a nesC program. Configurations are the second type of nesC component. They assemble components into larger abstractions.

In a nesC program, there are usually more configurations than modules. Except for low-level hardware abstractions, any given component is built on top of other components, which are encapsulated in configurations. For example, Figure 4.1 shows a routing stack (CollectionC) which depends on a single-hop packet layer (ActiveMessageC), which is itself a configuration. This single-hop configuration wires the actual protocol implementation module (e.g. setting header fields) to a raw packet layer on top of the radio (Link Layer). This raw packet layer is a configuration that wires the module which sends bytes out to the bus over which it sends bytes. The bus, in turn, is a configuration. These layers of encapsulation generally reach very low in the system.

Encapsulating an abstraction in a configuration means that it can be ready-to-use: all we need to do is wire to its provided interfaces. In contrast, if it were a module that uses and provides interfaces, then we'd need to wire up dependencies and requirements as well. For example, a radio stack can use a wide range of resources, including buses, timers, random number generators, cryptographic support, and hardware pins. This is the signature of just one module of the CC1000 (a wireless radio chip from TI used in the mica2 family motes) radio stack:

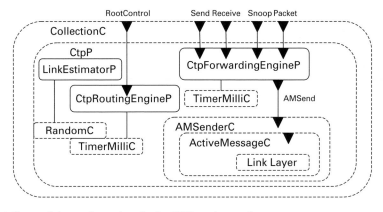

Figure 4.1 Some of the configurations in the CTP routing stack.

```
module CC1000CsmaP {
  provides {
    interface Init;
    interface SplitControl;
    interface CsmaControl;
    interface CsmaBackoff;
    interface LowPowerListening;
  }
  uses {
    interface Init as ByteRadioInit;
    interface StdControl as ByteRadioControl;
    interface ByteRadio;
    interface CC1000Control;
    interface CC1000Squelch;
    interface Random;
    interface Timer<TMilli> as WakeupTimer;
    interface BusyWait<TMicro, uint16_t>;
    interface ReadNow<uint16_t> as RssiNoiseFloor;
    interface ReadNow<uint16_t> as RssiCheckChannel;
    interface ReadNow<uint16_t> as RssiPulseCheck;
  }
```

Listing 4.1 Signature of part of the CC1000 radio stack

Rather than expecting a programmer to connect the stack up to all of these things, the entire stack can be encapsulated in a single component (CC1000ActiveMessageC). This component connects all of the radio's subcomponents so their dependencies needs are met.

Configurations also need to *export* interfaces. This kind of wiring, rather than connect a provider and a user, maps one name to another. Interface exports allow configurations to manage the namespace in a nesC program and act like modules in terms of providing or using interfaces. Managing the nesC namespace, and correspondingly, interface exports, is one of the most challenging aspects of nesC programming, and so this chapter goes over it in detail with many examples.

4.1 Configurations

Chapter 3 gave a brief introduction to configurations, showing examples of top-level configurations. Because top-level configurations wire components together into a completed application, they neither provide nor use interfaces. This lack of entries in the component signature is uncommon: most configurations are systems or abstractions that other components can use.

Let's revisit the syntax of a configuration. The first block of a configuration is its signature, which states that interfaces that the component uses or provides. The second block of a configuration is its implementation, which names components and wires them together. The implementation block of a configuration names the components it is wiring with the **components** keyword. Any number of component names besides zero can follow **components**, and their order does not matter. A configuration can have multiple **components** statements.

A configuration wires component interfaces using the three wiring operators: ->, <-, and =. This chapter discusses the -> and <- introduced in Chapter 3 in greater depth and introduces the = operator, which allows configurations to provide and use interfaces.

4.1.1 The −> and <− operators

The -> operators connect providers and users, binding callers and callees. Let's return to the PowerupToggle application and step through how its wiring works. The module PowerupToggleC (Listing 3.16, page 30) **uses** the Leds interface. The configuration PowerupToggleAppC wires PowerupToggleC.Leds to LedsC.Leds:

```
configuration PowerupToggleAppC {}
implementation {
  components MainC, LedsC, PowerupToggleC;

  PowerupToggleC.Boot -> MainC.Boot;
  PowerupToggleC.Leds -> LedsC.Leds;
}
```

Listing 4.2 The PowerupToggleAppC configuration revisited

In turn, LedsC maps LedsC.Leds to LedsP.Leds (see Section 4.1.2). The nesC wiring statements in PowerupToggleAppC and LedsC connect the Leds.led0Toggle (a name local to PowerupToggleC) command used in PowerupToggleC to LedsP's identically named Leds.led0Toggle command implementation. When PowerupToggleC calls its Leds.led0Toggle command, it actually calls LedsP's Leds.led0Toggle command implementation. The same is true for other calls of the Leds interface, such as Leds.led1On. The PowerupToggleAppC and LedsC configurations provide a mapping between the local namespaces of the PowerupToggleC and LedsP modules. The C code generated (as an intermediate step) by the nesC compiler looks something like this:

```
void LedsP_Leds_led0Toggle() { ... }

void PowerupToggleC_Leds_led0Toggle() {
  LedsP_Leds_led0Toggle();
}

...

void PowerupToggleC_Boot_booted() {
  while(1) {
    call PowerupToggleC_Leds_led0Toggle();
```

```
        call PowerupToggleC_Leds_led1Toggle();
        call PowerupToggleC_Leds_led2Toggle();
      }
    }
```

Listing 4.3 C code generated from the PowerupToggleAppC configuration

All of these levels of indirection could add significant overhead. Toggling an LED takes two function calls. In practice, however, the nesC compiler toolchain cuts out all of this overhead through extensive inlining. All of those function calls collapse and the LED toggling logic is embedded in the while loop.

4.1.2 The = operator

The -> and <- operators connect concrete providers and users that the configuration names through the **components** keyword. A direct wiring (a -> or <-) always goes from a user to a provider, resolving command and event call paths.

From the perspective of someone using a component, it shouldn't be relevant whether the component is a module or a configuration. Just like modules, configurations can provide and use interfaces. But as they have no code, these interfaces must be defined in terms of other components. The only way a configuration can provide or use an interface is to do so by proxy: it renames another component's implementation as its own. Configurations achieve this with the wiring operator, =. The = operator connects the interfaces in a configuration's signature to interfaces in components named in its **components** statements.

For example, this is the implementation of the configuration LedsC, which provides the Leds interface by *exporting* the interface provided by LedsP:

```
configuration LedsC {
  provides interface Leds;
}
implementation {
  components LedsP, PlatformLedsC;
  Leds = LedsP.Leds;

  LedsP.Init <- PlatformLedsC.Init;
  LedsP.Led0 -> PlatformLedsC.Led0;
  LedsP.Led1 -> PlatformLedsC.Led1;
  LedsP.Led2 -> PlatformLedsC.Led2;
}
```

Listing 4.4 The LedsC configuration

LedsC is a simple example of a configuration that connects a few small building blocks into a larger, more useful abstraction. LedsP is a simple module whose code implements the Leds interface commands by manipulating underlying digital IO lines

4.1 Configurations

accessed by the GeneralIO interface. PlatformLedsC provides three such IO lines, and uses an Init interface which is wired to LedsP.Init to initialize LedsP at system boot time.

From a programming standpoint, the configuration operators have two very different purposes. The -> and <- operators combine existing components, completing existing signatures. The = operator defines how a configuration's interfaces are implemented. Like a module, a configuration is an abstraction defined by a signature. Modules directly implement their functions (events from used interfaces, commands from provided interfaces). Configurations delegate implementations to other components using the = operator. For example, LedsC delegates the implementation of the Leds interface to LedsP.

4.1.3 Namespace management

We saw in Section 3.1.2 the use of **as** to manage the names in a component signature, e.g., MainC has

```
uses interface Init as SoftwareInit;
```

The **as** keyword can also be used within configurations. Because nesC components are in a global namespace, sometimes they have very long and descriptive names. For example, the lowest level (byte) SPI bus abstraction on the ATmega128 is HplAtm128SpiC, which means, "This is the hardware presentation layer component of the ATmega128 SPI bus." Typing that in a configuration is painful and not very easy to read. So, the slightly higher level abstraction, the configuration Atm128SpiC, names it like this:

```
component HplAtm128SpiC as HplSpi;
```

which makes the wiring significantly clearer. As was the case with interfaces, all **components** statements have an implicit use of **as**:

```
components MainC;
```

is just shorthand for

```
components MainC as MainC;
```

Another example of using **as** to clarify a configuration is found in CC2420ReceiveC, the receive path of the CC2420 radio. This configuration wires packet logic to things like interrupts and status pins:

```
configuration CC2420ReceiveC {...}
implementation {
  components CC2420ReceiveP;
  components new CC2420SpiC() as Spi;
```

```
  components HplCC2420PinsC as Pins;
  components HplCC2420InterruptsC as InterruptsC;
  // rest of the implementation elided
}
```

Listing 4.5 CC2420ReceiveC's use of the **as** keyword

This example shows a common use of **as**: to name the result of instantiating a generic component (CC2420SpiC). In fact, the use of **as** is required if the same generic component is instantiated twice in the same configuration, as in the BlinkAppC example we saw earlier:

```
configuration BlinkAppC {}
implementation {
  components MainC, BlinkC, LedsC;
  components new TimerMilliC() as Timer0;
  components new TimerMilliC() as Timer1;
  components new TimerMilliC() as Timer2;

  BlinkC.Timer0 -> Timer0.Timer;
  BlinkC.Timer1 -> Timer1.Timer;
  BlinkC.Timer2 -> Timer2.Timer;
  ...
}
```

Listing 4.6 Naming generic component instances

Without **as**, there would be no way to distinguish the three timers.

The **as** keyword makes configurations more readable and comprehensible. Because there is a flat component namespace, some components have long and complex names which can be easily summarized. Additionally, by using the **as** keyword, you create a level of indirection. For example, if a configuration uses the **as** keyword to rename a component, then changing the component only requires changing that one line. Without the keyword, you have to change every place it's named in the configuration.

4.1.4 Wiring rules

If a component uses the **as** keyword to change the name of an interface, then wiring must use this name. Returning to MainC and LedsP as examples:

4.1 Configurations

```
configuration MainC {                module LedsP {
  provides interface Boot;             provides {
  uses interface Init as SoftwareInit;   interface Init;
}                                        interface Leds;
                                       }
                                       uses {
                                         interface GeneralIO as Led0;
                                         interface GeneralIO as Led1;
                                         interface GeneralIO as Led2;
                                       }
```

Listing 4.7 MainC and LedsP

This wiring is valid:

```
MainC.SoftwareInit -> LedsP.Init;
```

but this is not:

```
MainC.Init -> LedsP.Init;
```

Finally, a configuration must name a component before it wires it. For example, this is a valid rewriting of PowerupToggleAppC:

```
configuration PowerupToggleAppC {}
implementation {
  components MainC, PowerupToggleC;
  PowerupToggleC.Boot -> MainC.Boot;

  component LedsC;
  PowerupToggleC.Leds -> LedsC.Leds;
}
```

Listing 4.8 Valid alternate of PowerupToggleAppC

while this version is invalid:

```
configuration PowerupToggleAppC {}
implementation {
  components PowerupToggleC;
  PowerupToggleC.Boot -> MainC.Boot; // Invalid: MainC not named yet

  component LedsC, MainC;
  PowerupToggleC.Leds -> LedsC.Leds;
}
```

Listing 4.9 Invalid alternate of PowerupToggleAppC

4.1.5 Wiring shortcuts

All interfaces have a *type*: it is not possible to wire, e.g., a Leds interface to a Boot interface, or a Read<uint8_t> to a Read<int16_t>. As a result, when wiring you can sometimes elide one of the interface names. For instance, you can change PowerupToggleAppC's wiring section to:

```
components MainC, LedsC, PowerupToggleC;
PowerupToggleC.Boot -> MainC;
PowerupToggleC -> LedsC.Leds;
```

On the left side, PowerupToggleC.Boot is an instance of the Boot interface. On the right side is MainC, without an interface name. Because MainC only provides one instance of the Boot interface, nesC assumes that this is the one you mean. Similarly, the wiring of component PowerupToggleC to interface Leds of LedsC must be specifying PowerupToggleC's Leds interface. So the two wiring statements are equivalent to:

```
PowerupToggleC.Boot -> MainC.Boot;
PowerupToggleC.Leds -> LedsC.Leds;
```

These shortcuts also apply to export wirings:

```
configuration LedsC {
  provides interface Leds;
}
implementation {
  components LedsP;
  Leds = LedsP;
  ...
}
```

Listing 4.10 LedsC revisited

The Leds interface of LedsC is wired to component LedsP, which implicitly resolves to the Leds interface of LedsP. However, the following code is invalid, as you can't omit the name of the exported interface:

```
= LedsP.Init;
```

Wiring shortcuts are based on interface types and whether the interface is provided or used, not names. The BlinkC module has the following signature:

```
module BlinkC
{
  uses interface Timer<TMilli> as Timer0;
  uses interface Timer<TMilli> as Timer1;
```

```
  uses interface Timer<TMilli> as Timer2;
  uses interface Leds;
  uses interface Boot;
}
```

Listing 4.11 BlinkC signature

If nesC sees the code

```
components new TimerMilliC() as Timer0;
BlinkC.Timer0 -> Timer0;
```

it knows that BlinkC.Timer0 is a used Timer<TMilli> interface. It searches the signature of Timer0 for a provided instance of the same interface, and finds an unambiguous match, so it can correctly wire the two even though a shortcut is used. If the wiring is ambiguous, then nesC cannot complete it, and it will report an error. For example,

```
BlinkC -> Timer0.Timer;
```

is ambiguous because there are three interfaces that BlinkC uses which could wire to Timer0.Timer.

4.2 Building abstractions

RandomC defines the standard TinyOS random number generator, a simple and heavily used abstraction:

```
configuration RandomC {
  provides interface Init;
  provides interface ParameterInit<uint16_t> as SeedInit;
  provides interface Random;
}
implementation {
  components RandomMlcgC;

  Init = RandomMlcgC;
  SeedInit = RandomMlcgC;
  Random = RandomMlcgC;
}
```

Listing 4.12 The RandomC configuration

Mlcg stands for "multiplicative linear congruential generator." In the default case, RandomC is a wrapper around RandomMlcgC. There's another implementation, RandomLfsrC, that is about twice as fast but whose random numbers are not as good. Platforms or applications that need to use RandomLfsrC can redefine RandomC to encapsulate RandomLfsrC instead.

4.2.1 Component naming

As we saw earlier, TinyOS makes the distinction between components that end in C (an externally usable abstraction) and P (an internal implementation). Once you have written the signature for an externally usable component, changing it is very hard: any number of other components might depend on it, and changing it will cause compilation errors. In contrast, because an internal implementation is only wired to by higher-level configurations within that software abstraction, their signatures are much more flexible. For instance, changing the signature of LedsC would break almost all TinyOS code, but an internal change to LedsP (and changing its wiring in LedsC) should not be apparent to the user.

> **Programming Hint 11** IF A COMPONENT IS A USABLE ABSTRACTION BY ITSELF, ITS NAME SHOULD END WITH C. IF IT IS INTENDED TO BE AN INTERNAL AND PRIVATE PART OF A LARGER ABSTRACTION, ITS NAME SHOULD END WITH P. NEVER WIRE TO P COMPONENTS FROM OUTSIDE YOUR PACKAGE (DIRECTORY).

Let's look at a complete (but very simple) example of how all of these issues are resolved in RandomC.

As shown above, RandomC maps to a specific implementation, RandomMlcgC. RandomMlcgC is a configuration whose main purpose is to expose the random number implementation (RandomMlcgP) and wire it to the boot sequence (we discuss component initialization further below):

```
configuration RandomMlcgC {
  provides interface Init;
  provides interface ParameterInit<uint16_t> as SeedInit;
  provides interface Random;
}
implementation {
  components RandomMlcgP, MainC;

  MainC.SoftwareInit -> RandomMlcgP; // Auto-initialize
  Init = RandomMlcgP; // Allow for re-initialization

  SeedInit = RandomMlcgP;
  Random = RandomMlcgP;
}
```

Listing 4.13 The RandomMlcgC signature

RandomMlcgP is a software implementation. A platform that has a hardware random number generator could have a different RandomMlcgP. Because this different implementation might have a different signature – e.g. it might require accessing registers through an abstraction layer – it might also require a different RandomMlcgC, to resolve these dependencies and present a complete abstraction.

In short, the configuration RandomC maps the standard number generator to a specific algorithm, RandomMlcgC. The configuration RandomMlcgC encapsulates a specific implementation as a complete abstraction. RandomMlcgP is a software implementation of the multiplicative linear congruential generator. Similarly, there is also a RandomLfsrC, which is a linear feed shift register random number generator. Just like RandomMlcgC, RandomLfsrC is a configuration that exports the interfaces of RandomLfsrP and wires it to the boot sequence. This hierarchy of names means that a system can wire to a specific random number generator if it cares which one it uses, or wire to the general one that TinyOS provides (RandomC). An application can change what the default random number generator is by defining its own RandomC, which maps to a different algorithm. The TinyOS developers can change and improve RandomMlcgP without worrying about breaking application code. Finally, a platform implementer can replace RandomMlcgC with an optimized, platform-specific version as long as it has the same signature as RandomMlcgC.

4.2.2 Component initialization

RandomMlcgP provides the Init interface to seed the random number generator with the node's identifier (TOS_NODE_ID). TOS_NODE_ID is a global constant in the TinyOS program. The TinyOS tutorials describe how to set it when you install a program on a mote. SeedInit starts the generator with a specific seed:

```
module RandomMlcgP {
  provides interface Init;
  provides interface ParameterInit<uint16_t> as SeedInit;
  provides interface Random;
}
implementation {
  uint32_t seed;

  /* Initialize the seed from the ID of the node */
  command error_t Init.init() {
    seed = (uint32_t)(TOS_NODE_ID + 1);
    return SUCCESS;
  }

  /* Initialize with 16-bit seed */
  command error_t SeedInit.init(uint16_t s) {
    seed = (uint32_t)(s + 1);
    return SUCCESS;
  }
  ...
}
```

Listing 4.14 Seed initialization in RandomMlcgP

RandomMlcgC wires RandomMlcgP.Init in two different ways:

```
MainC.SoftwareInit -> RandomMlcgP;  // Auto-initialize
Init = RandomMlcgP;  // Allow for re-initialization
```

In the first wiring, RandomMlcgC wires RandomMlcgP.Init to the TinyOS boot sequence (MainC). When TinyOS boots, it calls MainC's SoftwareInit.init, and so it calls RandomMlcgP's Init.init. In the second, it equates its own RandomMlcgC.Init with RandomMlcgP.Init. If a component calls RandomMlcgC's Init.init, it actually calls RandomMlcgP's Init.init.

The first wiring makes sure that RandomMlcgP has been properly seeded before an application starts. If the application reseeds it by calling RandomMlcgC.Init or RandomMlcgC.SeedInit, no harm is done. But by wiring to MainC, RandomMlcgC makes sure that an application (or protocol, or system) doesn't have to remember to initialize RandomMlcgC.

This technique – "auto-wiring" initialization – is used in many TinyOS abstractions. One very common bug in TinyOS 1.x was to forget to initialize. This usually happens because many components might be initializing the same component. This approach is wasteful, but since initialization only happens once, it's not a huge issue. The bigger issue is that a component often relies on someone else initializing. For example, imagine two radio stacks, A and B. A initializes the timer system, B does not. A programmer writes an application using radio stack A and forgets to initialize the timer system. Because radio stack A does, everything works fine. The programmer then decides to switch to radio stack B, and nothing works: neither the application nor the stack initialize the timers, and so the system just hangs.

The init commands of components using auto-wired initialization are called in an arbitrary order (see Section 4.4 below). For software initialization – setting fields, etc. – this generally doesn't matter (Init is not supposed to call anything besides another Init). Hardware initialization is a much trickier problem, and is generally handled on a per-platform basis. Refer to TEP 107 [14] for more details.

Programming Hint 12 AUTO-WIRE INIT TO MAINC IN THE TOP-LEVEL CONFIGURATION OF A SOFTWARE ABSTRACTION.

4.3 Component layering

RandomC shows how even a simple abstraction, such as a random number generator, can have several layers within it. On one hand, this programming style makes components small and reusable: for people who know them well, it makes building new systems the simple matter of combining existing building blocks and maybe adding a bit of new logic. On the other hand, it means getting to the bottom of a component – figuring out what it actually does – can be really frustrating to someone unfamiliar with the code. RandomC is a good example of this layering. A programmer, wanting to find the implementation (module code) of the random number generator, has to look at three files. RandomC points to RandomMlcgC, which points to RandomMlcgP.

4.3.1 Extensibility

In practice, these layers of indirection are not particular to nesC: any system with complex libraries has them. The difference is how those layers are written. In C code, you might see a series of functions that are simple wrappers (e.g. printf as a call to fprintf passing stdout as the file stream): you traverse executable code. In nesC code you traverse namespace mappings through configurations. The first few times you try to get to the bottom of a function, this difference can be a bit confusing, but it doesn't take very long to get used to it. You can also use the nesC's nesdoc tool to navigate through an application's wiring and find where an interface is ultimately implemented – see the TinyOS tutorials for more information.

Just as in other languages, especially object-oriented ones, nesC software design has a tension between how simple the implementations are and how easy it is to use or reuse them. The C++ standard template library is infamous for this: it is amazingly powerful, general, and efficient, but understanding the implementations is non-trivial. Putting all of a system's functionality in a single module means a user doesn't need to track down call chains. But it means that modifying the system requires replacing or copying that entire file. This leads to code replication, which greatly increases the lifetime of software bugs (if there's a bug in the copied code, you need to change all of the copies, and almost always forget one). Therefore, a lot of the core TinyOS abstractions, such as timers and communication, tend to be composed of many small components. This makes the systems more difficult to understand at a first read, but easy to tweak, replace, or modify once you do.

4.3.2 Hardware specificity

Furthermore, there's a tension between high-level APIs that are hardware independent and APIs that enable a program to take advantage of all of the hardware capabilities. For example, some but not all radios have multiple channels (frequencies). Some systems may want to take advantage of this functionality, and are willing to assume they are using a particular radio chip in order to do so. Other systems may want to remain hardware independent.

TinyOS uses component layering to make this possible. Let's look at packet communication as an example. The top-level component a platform needs to provide is called ActiveMessageC and is hardware independent.[1] ActiveMessageC is almost always a configuration that layers on top of a chip-specific component that provides additional interfaces. For example, this is the implementation of ActiveMessageC for the micaz mote, which has a CC2420 radio chip:

```
configuration ActiveMessageC {
  provides {
    interface SplitControl;
```

---

[1] TinyOS uses "active messages" [29], often abbreviated to AM, as its networking abstraction (see Section 6.3).

```
      interface AMSend[uint8_t id];
      interface Receive[uint8_t id];
      interface Receive as Snoop[uint8_t id];

      interface Packet;
      interface AMPacket;
      interface PacketAcknowledgements;
  }
}
implementation {
  components CC2420ActiveMessageC as AM;

  SplitControl = AM;
  AMSend       = AM;
  Receive      = AM.Receive;
  Snoop        = AM.Snoop;
  Packet       = AM;
  AMPacket     = AM;
  PacketAcknowledgements = AM;
}
```

Listing 4.15 ActiveMessageC for the CC2420

Don't worry about the interfaces with brackets ([and]); they're called parameterized interfaces and are an advanced nesC topic covered in Chapter 9. ActiveMessageC merely exports a bunch of interfaces from CC2420ActiveMessageC: interfaces for sending packets, receiving packets, and accessing packet fields (the AMPacket and Packet interfaces). But if you look at the CC2420ActiveMessageC component, it provides several interfaces which the micaz's ActiveMessageC *does not* export:

```
configuration CC2420ActiveMessageC {
  provides {
    interface SplitControl;
    interface AMSend[am_id_t id];
    interface Receive[am_id_t id];
    interface Receive as Snoop[am_id_t id];
    interface AMPacket;
    interface Packet;
    interface CC2420Packet;
    interface PacketAcknowledgements;
    interface RadioBackoff[am_id_t amId];
    interface LowPowerListening;
    interface PacketLink;
  }
}
```

Listing 4.16 The signature of CC2420ActiveMessageC

The hardware-independent ActiveMessageC doesn't export CC2420Packet, Radio-Backoff, LowPowerListening, or PacketLink. What ActiveMessageC does is a very common use of configurations. ActiveMessageC adds no functionality: all it does is export CC2420ActiveMessageC's interfaces and therefore give them different, alternative names in the global namespace. Calling CC2420ActiveMessageC's AMSend.send is the same as calling ActiveMessageC's AMSend.send. The difference lies in the kinds of assumptions the caller can make. For example, it's possible that if you wired a component to ActiveMessageC.AMSend and CC2420ActiveMessageC.CC2420Packet, you might run into serious problems. It might be that a platform has two radios, one of which is a CC2420, but ActiveMessageC doesn't refer to the CC2420. In this case, the component would access protocol fields that don't exist, and chaos is guaranteed.

Programming Hint 13 WHEN USING LAYERED ABSTRACTIONS, COMPONENTS SHOULD NOT WIRE ACROSS MULTIPLE ABSTRACTION LAYERS: THEY SHOULD WIRE TO A SINGLE LAYER.

4.4 Multiple wirings

Not all wirings are one-to-one. For example, this is part of the component CC2420TransmitC, a configuration that encapsulates the transmit path of the CC2420 radio:

```
configuration CC2420TransmitC {
  provides interface Init;
  provides interface AsyncControl;
  provides interface CC2420Transmit;
  provides interface CsmaBackoff;
  provides interface RadioTimeStamping;
}
implementation {
  components CC2420TransmitP;
  components AlarmMultiplexC as Alarm;
  Init = Alarm;
  Init = CC2420TransmitP;
  // further wirings elided
}
```

Listing 4.17 Fan-out on CC2420TransmitC's Init

This wiring means that CC2420TransmitC.Init maps both to Alarm.Init and CC2420TransmitP.Init. What does that mean? There certainly isn't any analogue in C-like languages. In nesC, a multiple-wiring like this means that when a component calls CC2420TransmitC's Init.init, it calls both Alarm's Init.init and CC2420TransmitP's Init.init. The order of the two calls is not defined.

4.4.1 Fan-in and fan-out

This ability to multiply wire might seem strange. In this case, you have a single call point, CC2420TransmitC's Init.init, which fans-out to two callees. There are also fan-ins, which are really just a fancy name for "multiple people call the same function." But the similarity of the names "fan-in" and "fan-out" is important, as nesC interfaces are bidirectional. For example, coming from C, wiring two components to RandomC.Random doesn't seem strange: two different components might need to generate random numbers. In this case, as Random only has commands, all of the functions are fan-in. There are multiple callers for a single callee, just like a library function.

But as nesC interfaces are bidirectional, if there is fan-in on the command of an interface, then when that component signals an event on the interface, there is fan-out (multiple callees). Take, for example, the power control interfaces, StdControl and SplitControl. StdControl is single-phase: it only has commands. SplitControl, as its name suggests, is split-phase: the commands have completion events:

```
interface StdControl {
  command error_t start();
  command error_t stop();
}
interface SplitControl {
  command error_t start();
  event void startDone(error_t error);

  command error_t stop();
  event void stopDone(error_t error);
}
```

Listing 4.18 StdControl and SplitControl initialization interfaces

With StdControl, the service is started (stopped) by the time start (stop) returns, while with SplitControl, the service is only guaranteed to have started when startDone is signaled.

In this wiring, there is fan-in on StdControl:

```
components A, B, C;
A.StdControl -> C;
B.StdControl -> C;
```

Then either A or B can call StdControl to start or stop C. However, in this wiring, there are also completion events, hence both fan-in and fan-out on SplitControl:

```
components A, B, C;
A.SplitControl -> C;
B.SplitControl -> C;
```

Either A or B can call SplitControl.start. When C issues the SplitControl.startDone event, though, both of them are wired to it, so both A's SplitControl.startDone and B's

SplitControl.startDone are called. The implementation has no way of determining which called the start command.[2]

In summary, interfaces are not a one-to-one relationship. Instead, they are an *n*-to-*k* relationship, where *n* is the number of users and *k* is the number of providers. Any provider signaling will invoke the event handler on all *n* users, and any user calling a command will invoke the command on all *k* providers.

4.4.2 Uses of multiple wiring

In practice, multiple wirings allow an implementation to be independent of the number of components it depends on. Remember that MainC (Listing 3.8, page 26) abstracts the boot sequence as two interfaces, SoftwareInit and Boot. MainC calls SoftwareInit.init when booting so that software components can be initialized before execution begins. MainC then signals Boot.booted once the entire boot sequence is over. Many components need initialization. For example, in the very simple application RadioCountToLeds, there are ten components wired to MainC.SoftwareInit. Rather than use many Init interfaces and call them in some order, MainC just calls SoftwareInit once and this call fans-out to all of the components that have wired to it.

Anecdote: Historically, multiple wirings come from the idea that TinyOS components can be thought of as hardware chips. In this model, an interface is a set of pins on the chip. The term wiring comes from this idea: connecting the pins on one chip to those of another. In hardware, though, you can easily connect N pins together. For example, a given general-purpose IO pin on a chip might have multiple possible triggers, or a bus have multiple end devices that are controlled with chip select pins. It turns out that taking this metaphor literally has several issues. When TinyOS moved to nesC, these problems were done away with. Specifically, consider this configuration:

```
configuration A {
  uses interface StdControl;
}

configuration B {
  provides interface StdControl;
  uses interface StdControl as SubControl; // Called in StdControl
}

configuration C {
  provides interface StdControl;
}

A.StdControl -> B.StdControl;
```

---

[2] There are ways to disambiguate this, through parameterized interfaces, which are covered in Chapter 9.

```
A.StdControl  ->  C.StdControl;
B.SubControl  ->  C.StdControl;
```

Listing 4.19 Why the metaphor of "wires" is only a metaphor

If you take the multiple wiring metaphor literally, then the wiring of B to C joins it with the wiring of A to B and C. That is, they all form a single "wire." The problem is that under this interpretation, B's call to C is the same wire as A's call to B. B enters an infinite recursion loop, as it calls SubControl, which calls B.StdControl, which calls SubControl, and so on and so on. Therefore, nesC does not take the metaphor literally. Instead, the wirings from one interface to another are considered separately. So the code

```
A.StdControl  ->  B.StdControl;
A.StdControl  ->  C.StdControl;
B.SubControl  ->  C.StdControl;
```

makes A's calls to StdControl.start call B and C, and B's calls to SubControl.start call C only.

4.4.3 Combine functions

Fan-out raises an interesting question: if

```
call SoftwareInit.init()
```

actually calls ten different functions, then what is its return value?

nesC provides the mechanism of *combine functions* to specify the return value. A data type can have an associated combine function. Because a fan-out always involves calling N functions with identical signatures, the N results can be combined to a single value by $N - 1$ calls to a combine function, each taking a pair of values of the result type. When nesC compiles the application, it auto-generates a fan-out function which applies the combine function.

For example, TinyOS's error type, error_t is a very common result type. Its combine function is ecombine:

```
error_t ecombine(error_t e1, error_t e2) {
    return (e1 == e2) ? e1 : FAIL;
}
```

Listing 4.20 The combine function for error_t

If both calls return the same value, ecombine returns that value. Otherwise, it returns FAIL. Thus combining two SUCCESS values returns SUCCESS, combining two identical Exxx error codes returns that code, and all other combinations return FAIL.

4.4 Multiple wirings

This combine function is bound to error_t with a nesC attribute (attributes are covered in Section 8.4):

```
typedef uint8_t error_t @combine("ecombine");
```

When asked to compile the following configuration

```
configuration InitExample {}
implementation {
  components MainC;
  components AppA, AppB, AppC;

  MainC.SoftwareInit -> AppA;
  MainC.SoftwareInit -> AppB;
  MainC.SoftwareInit -> AppC;
}
```

Listing 4.21 Fan-out on SoftwareInit

the nesC compiler will generate something like the following code:[3]

```
error_t MainC__SoftwareInit__init() {
  error_t result;
  result = AppA__SoftwareInit__init();
  result = ecombine(result, AppB__SoftwareInit__init());
  result = ecombine(result, AppC__SoftwareInit__init());
  return result;
}
```

Listing 4.22 Resulting code from fan-out on SoftwareInit

Combine functions should be associative and commutative, to ensure that the result of a fan-out call does not depend on the order in which the commands (or event) are executed (in the case of fan-out, this order is picked by the compiler).

Some return values don't have combine functions, either due to programmer oversight or the semantics of the data type. Examples of the latter include things like data pointers: if both calls return a pointer, say, to a packet, there isn't a clear way to combine them into a single pointer. If your program has fan-out on a call whose return value can't be combined, the nesC compiler will issue a warning along the lines of

"calls to Receive.receive in CC2420ActiveMessageP are uncombined"

or

"calls to Receive.receive in CC2420ActiveMessageP fan out, but there is no combine function specified for the return value."

---

[3] The nesC compiler actually compiles to C, which it then passes to a native C compiler. Generally, it uses _ as the delimiter between component, interface, and function names. By not allowing _ in the middle of names, the nesC compiler enforces component encapsulation (there's no way to call a function with a _ from within nesC and break the component boundaries).

Programming Hint 14 NEVER IGNORE COMBINE WARNINGS.

4.5 Generics versus singletons

Configurations can be generics, just as modules can. For example, the standard TinyOS abstraction for sending packets is the component AMSenderC, a generic that takes a single parameter, the AM (active message) type of packet to be sent:

```
generic configuration AMSenderC(am_id_t AMId) {
  provides {
    interface AMSend;
    interface Packet;
    interface AMPacket;
    interface PacketAcknowledgements as Acks;
  }
}
```

Listing 4.23 AMSenderC signature

AM types allow TinyOS applications to send multiple types of packets: they are somewhat like ports in UDP/TCP sockets. AM types are 8 bits. There is also a generic receiving component, AMReceiverC, which takes an AM type as a parameter. AM types allow multiple components and subsystems to receive and send packets without having to figure out whose packets are whose, unless two try to use the same AM type.

This raises the question: why aren't all components generics? There are some low-level components that inherently can't be generics, such as those that provide direct access to hardware. Because these components actually represent physical resources (registers, buses, pins, etc.), there can only be one. There are very few of these low-level components though. More commonly, singletons are used to provide a global name which many components can use. To provide better insight on why this is important, let's step through what generics are and how they work.

4.5.1 Generic components, revisited

Generic components allow many systems to use independent copies of a single implementation. For example, a large application might include multiple network services, each of which uses one or more instances of AMSenderC to send packets. Because all share a common implementation, bugs can be fixed and optimizations can be applied in a single place, and over time they will all benefit from improvements.

While a generic component has a global name, such as AMSenderC, each instance has a local name. Two separate references to new AMSenderC are two separate components. Take, for example, the application configuration RadioCountToLedsAppC:

4.5 Generics versus singletons

```
configuration RadioCountToLedsAppC {}
implementation {
  components MainC, RadioCountToLedsC as App, LedsC;
  components new AMSenderC(AM_RADIO_COUNT_MSG);
  components new AMReceiverC(AM_RADIO_COUNT_MSG);
  components new TimerMilliC();
  components ActiveMessageC;

  App.Boot -> MainC.Boot;

  App.Receive -> AMReceiverC;
  App.AMSend -> AMSenderC;
  App.AMControl -> ActiveMessageC;
  App.Leds -> LedsC;
  App.MilliTimer -> TimerMilliC;
  App.Packet -> AMSenderC;
}
```

Listing 4.24 RadioCountToLedsAppC

Because RadioCountToLedsAppC has to instantiate a new AMSenderC, that instance is private to RadioCountToLedsAppC. No other component can name that AMSenderC's interfaces or access its functionality. For example, RadioCountToLedsAppC does not wire AMSenderC's PacketAcknowledgements interface, either through direct wiring via -> or through an export via =. Since this AMSenderC is private to RadioCountToLedsAppC, that means no component can wire to its PacketAcknowledgements, because they cannot name it.

While generics enable components to reuse implementations, sharing a generic among multiple components requires a little bit of work. For example, let's say you want to share a pool of packet buffers between two components. TinyOS has a component, PoolC, which encapsulates a fixed-size pool of objects which components can dynamically allocate and free. PoolC is a generic configuration that takes two parameters, the type of memory object in the pool and how many there are:

```
generic configuration PoolC(typedef pool_t, uint8_t POOL_SIZE) {
  provides interface Pool<pool_t>;
}
```

Listing 4.25 PoolC

How do we share this pool between two different components? One way would be to write a configuration that wires both of them to a new instance:

```
components A, B, new PoolC(message_t, 8);
A.Pool -> PoolC;
B.Pool -> PoolC;
```

But what if we don't even know which two want to share it? For example, it might be that we just want to have a shared packet pool, which any number of components can use. Making a generic's interfaces accessible across a program requires giving it a global name.

4.5.2 Singleton components, revisited

Unlike generics, singleton components introduce a global name for a component instance that any other component can reference. So, to follow the previous example, one easy way to have a pool that any component can use would be to write it as a singleton component, say PacketPoolC. But we'd like to be able to do this without copying all of the pool code. It turns out that doing so is very easy: you just give a generic instance a global name by wrapping it up in a singleton. For example, here's the implementation of PacketPoolC, assuming you want an 8-packet pool:

```
configuration PacketPoolC {
  provides interface Pool<message_t>
}
implementation {
  components new PoolC(message_t, 8);
  Pool = PoolC;
}
```

Listing 4.26 Exposing a generic component instance as a singleton

All PacketPoolC does is instantiate an instance of PoolC that has 8 message_t structures, then exports the Pool interface as its own. Now, any component that wants to access this shared packet pool can just wire to PacketPoolC.

While you can make singleton instances of generic components in this way, you can't make generic versions of singleton components. Singletons inherently have only a single copy of their code. Every component that wires to PacketPoolC wires to the same PacketPoolC: there is no way to create multiple copies of it. If you needed two packet pools, you could just make another singleton with a different name.

4.6 Exercises

1. Take the TinyOS demo application RadioCountToLeds and trace through its components to figure out which components are auto-wired to SoftwareInit. There might be more than you expect: the Telos platform, for example, has 10. Hint: you can use the nesdoc tool to simplify your task.
2. If you want multiple components to handle a received packet, you must either not allow buffer-swapping or must make a copy for all but one of the handlers. Write a component library that lets an application create a component supporting a non-swapping reception interface. Hint: you'll need to write a singleton wrapper.

5 Execution model

This chapter presents TinyOS's execution model, which is based on split-phase operations, run-to-completion tasks and interrupt handlers. Chapter 3 introduced components and modules, Chapter 4 introduced how to connect components together through wiring. This chapter goes into how these components execute, and how you can manage the concurrency between them in order to keep a system responsive. This chapter focuses on tasks, the basic concurrency mechanism in nesC and TinyOS. We defer discussion of concurrency issues relating to interrupt handlers and resource sharing to Chapter 11, as these typically only arise in very high-performance applications and low-level drivers.

5.1 Overview

As we saw in Section 3.4, all TinyOS I/O (and long-running) operations are split-phase, avoiding the need for threads and allowing TinyOS programs to execute on a single stack. In place of threads, all code in a TinyOS program is executed either by a *task* or an interrupt handler. A task is in effect a lightweight deferred procedure call: a task can be posted at anytime and posted tasks are executed later, one-at-a-time, by the TinyOS scheduler. Interrupts, in contrast can occur at any time, interrupting tasks or other interrupt handlers (except when interrupts are disabled).

While a task or interrupt handler is declared within a particular module, its execution may cross component boundaries when it calls a command or signals an event (Figure 5.1). As a result, it isn't always immediately clear whether a piece of code is only executed by tasks or if it can also be executed from an interrupt handler. Because code that can be executed by an interrupt handler has to be much more aware of concurrency issues (as it may be called at any time), nesC distinguishes between synchronous (sync) code that may only be executed by tasks and asynchronous (async) code that may be executed by both tasks and interrupt handlers, and requires that asynchronous commands or events be declared with the **async** keyword (both in interfaces and in the actual module implementation).

Writing components that implement async functions requires a few advanced nesC features, and tasks and synchronous code are sufficient for many applications so we defer further discussion of asynchronous code and interrupt handlers to Chapter 11.

Execution model

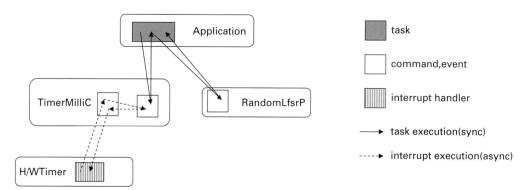

Figure 5.1 TinyOS execution model: both tasks (full lines) and interrupt handlers (dotted lines) cross component boundaries.

5.2 Tasks

A task is a simple deferred computation mechanism provided by nesC. Tasks have a return value of void, take no parameters (this keeps the task scheduler very simple) and are declared with the **task** keyword:

```
task void setupTask() {
  // task code
}
```

Using the **post** keyword schedules a task for later execution:

```
event void Boot.booted() {
  call Timer.startPeriodic(1024);
  post setupTask();
}
```

In the above example, once TinyOS completes the boot sequence (signals Boot. booted to all components), it will run setupTask.

Tasks are like any other module code. They can access variables, call commands, signal events, and invoke internal or global C functions. The post operator is like calling a function: it has a return value of error_t. Posting a task always returns SUCCESS unless the task is already pending. Put another way, a component cannot post multiple copies of the same task to run; however, once a task has started executing it may repost itself.

This is the core TinyOS scheduler loop that runs after the system boots. This function never returns. It executes tasks until the task queue is empty, then puts the microcontroller to sleep:

```
command void Scheduler.taskLoop() {
  for (;;) {
    uint8_t nextTask;
    atomic {
      while ((nextTask == popTask()) == NO_TASK) {
        call McuSleep.sleep();
      }
      signal TaskBasic.runTask[nextTask]();
    }
  }
}
```

Listing 5.1 The main TinyOS scheduling loop from SchedulerBasicP.nc

Don't worry about some of the advanced syntax in the loop (such as the []). In pseudocode, the function is:

```
run forever:
  while there are no tasks:
    sleep
  run next task
```

That's it. TinyOS sleeps until an interrupt wakes up the processor. If the interrupt handler posted one or more tasks, TinyOS runs tasks until there are no more left, then goes back to sleep. Since tasks are run one by one, they can't preempt one another, and the next task doesn't execute until the current one completes. This property greatly simplifies writing tasks, and hence all sync code, as you don't need any locks or other mechanisms to protect shared variables. A sync routine can assume it has complete control over the processor until it completes.

As an example, let's look at the BaseStation application. BaseStation is a UART/radio bridge: it forwards all radio packets it receives to the UART, and forwards all UART packets it receives to the radio. Because the radio and UART might have different throughputs, BaseStation introduces a send queue in each direction. When it receives a radio packet, it puts the packet on the UART send queue. When it receives a UART packet, it puts the packet on the radio send queue.

BaseStationP, the module that implements the BaseStation application, uses tasks to pull packets off the queues and send them. The receive handlers put packets on the queue and post the send task if the application is not already sending. Here is pseudocode of the logic:

```
on receive packet:
  if queue not full:
    put packet on queue
    if no send pending:
      post send task
```

Tasks cannot take parameters. If a component needs to pass data to a task, it has to do so by storing it in a component variable. For example, BaseStationP cannot pass a pointer to the message to send as a parameter to sendTask. Instead, sendTask must pull the packet off of the queue directly.

5.2.1 Task timing

TinyOS normally runs tasks in the same order they're posted (FIFO). A posted task won't run until all tasks posted before it complete. This means that tasks should be pretty short. If a component has a very long computation to do, it should break it up into multiple tasks. A task can post itself, as once it is running it is no longer on the queue.

It takes about 80 microcontroller clock cycles to post and execute a task on current mote platforms. Generally, keeping task run times to at most a few milliseconds is a good idea. Because tasks are run to completion, then a long-running task or large number of not-so-long-running tasks can introduce significant latency (tens of milliseconds) between a task post and its execution. This usually isn't a big deal with application-level components. But there are lower-level components, such as radio stacks, that use tasks. For example, if the packet reception rate is limited by how quickly the radio can post tasks to signal reception, then a latency of 10 ms will limit the system to 100 packets per second.

Consider these two cases. In both, there are five processing components and a radio stack. The mote processor runs at 8 MHz. Each processing component needs to do a lot of CPU work. In the first case, the processing components post tasks that run for 5 ms and repost themselves to continue the work. In the second case, the processing components post tasks that run for 500 μs and repost themselves to continue the work.

In the first case, the task posting overhead is 0.02%: 80 cycles overhead on 40 000 cycles of execution. In the second case, the task posting overhead is 0.2%: 80 cycles overhead on 4 000 cycles of execution. So the time to complete the executions isn't significantly different. However, consider the task queue latency. In the first case, when the radio stack posts a task to signal that a packet has been received, it expects to wait around 25 ms (5 processing tasks × 5 ms each), limiting the system to 40 packets per second. In the second case, when the radio stack posts the task, it expects to wait around 2.5 ms (5 processing tasks × 500 μs each), limiting the system to 400 packets per second. Because the task posting cost is so low, using lots of short running tasks improves the responsiveness of the system without introducing significant CPU overhead.

Of course, there's often a tradeoff between lots of short tasks and the amount of state you have to allocate in a component. For example, let's say you want to encrypt a chunk of data. If the encryption operation takes a while (e.g. 10 ms), then splitting it into multiple task executions would improve the overall system responsiveness. However, if you execute it in a single task, then you can allocate all of the state and scratch space you need on the stack. In contrast, splitting it across tasks would require keeping this state and scratch space in the component. There is no hard rule on this tradeoff. But generally, long-running tasks can cause other parts of the OS to perform poorly, so should be avoided when possible.

Programming Hint 15 KEEP TASKS SHORT.

5.2.2 Timing and event handlers

The need for tasks to be short directly affects how you implement components, in particular, event handlers. BaseStationP doesn't directly send packets in its receive event handlers; instead, it posts tasks to send the packets. It does this because the lower-level radio stack is signaling receive from within a task, presumably after a bit of computation. If the call to send takes a lot of cycles, then the lower-level radio component will not get a new buffer from the application until send completes. More generally, if the receive handler has significant computation in it, then the radio has to wait for that to complete before it has a buffer into which to receive the next packet.

While a single event handler may not be a big deal, an event handler may actually represent several software layers. For example, a networking component may handle a receive event, perform a small computation based on the packet, and then signal it to the next layer. Therefore, any given component may be just one part of a long chain of event handlers. For this reason, if a handler needs to perform significant computation, it is best to post a task. Doing so prevents a call chain from having multiple such handlers.

> **Programming Hint 16** IF AN EVENT HANDLER NEEDS TO MAKE POSSIBLY LONG-EXECUTING COMMAND CALLS, POST A TASK TO MAKE THE CALLS.

This is why BaseStationP uses tasks to send packets rather than do so directly in event handlers. While tasks may in theory have to wait a while before they run, in practice tasks tend to be very short, and so there is little latency between posting and execution.

5.3 Tasks and split-phase calls

Tasks do more than provide a way to maintain system responsiveness with a single stack. Tasks enable nesC programs to have a flexible hardware/software boundary: software components can behave similarly to hardware. Explaining why this is challenging requires a bit of a digression into how most peripherals, such as sensors and radios, work.

5.3.1 Hardware versus software

Split-phase calls represent how most peripherals work. Software issues commands to a device, and some time later the device indicates the operation is complete, typically with an interrupt. The device driver's interrupt handler signals the operation's completion event. While the device is busy, the processor can continue to issue instructions and do other useful work. Therefore, the command that starts the operation can return immediately and allow the program to continue.

The key point is that, for hardware implementations, the interrupt handler invokes driver code. But what happens if the implementation is purely software? For example, SineSensorC, which we saw in Chapter 3, is a purely software sensor that computes a sine value. As we want SineSensorC to be interchangeable with hardware sensors, it provides the same split-phase interface, Read. When a component calls the read comment, SineSensorC needs to signal readDone with the next reading.

5.3.2 Tasks and call loops

Let's return to Read, the interface most sensors provide for users to generate sensor readings. Read has a single command, read, and a single event, readDone. Let's imagine we have a sensor that's very noisy. To try to filter out some of that noise, an application needs a simple filter component that smooths the raw readings with an exponentially weighted moving average (EWMA):

```
module FilterMagC {
  provides interface StdControl;
  provides interface Read<uint16_t>;
  uses interface Timer<TMilli>;
  uses interface Read<uint16_t> as RawRead;
}
implementation {
  uint16_t filterVal = 0;
  uint16_t lastVal = 0;

  error_t StdControl.start() {
    return call Timer.startPeriodic(10);
  }
  command error_t StdControl.stop() {
    return call Timer.stop();
  }
  event void Timer.fired() {
    call RawRead.read();
  }
  event void RawRead.readDone(error_t err, uint16_t val) {
    if (err == SUCCESS) {
      lastVal = val;
      filterVal *= 9;
      filterVal /= 10;
      filterVal += lastVal / 10;
    }
  }

  command error_t Read.read() {
    signal Read.readDone(SUCCESS, filterVal);
  }
}
```

Listing 5.2 A troublesome implementation of a magnetometer sensor

The driver samples the magnetometer every 10 ms and applies an EWMA to those values. When the application samples this filtered value by calling Read.read, FilterMagC just signals Read.readDone with the cached, filtered value.

5.3 Tasks and split-phase calls

On one hand, this approach is very simple and fast. On the other, it can lead to significant problems with the stack. Imagine, for example, a component, FastSamplerC, that wants to sample a sensor many times quickly (acquire a high-frequency signal). It does this by calling Read.read in its Read.readDone handler:

```
event void Read.readDone(error_t err, uint16_t val) {
  buffer[index] = val;
  index++;
  if (index < BUFFER_SIZE) {
    call Read.read();
  }
}
```

Listing 5.3 Signal handler that can lead to an infinite loop

If, for some reason, an application wired FastSamplerC to FilterMagC, then there would be a long call loop between read and readDone. If the compiler can't optimize the function calls away, this will cause the stack to grow significantly. Given that motes often have limited RAM and no hardware memory protection, exploding the stack like this can corrupt data memory and cause the program to crash.

Programming Hint 17 DON'T SIGNAL EVENTS FROM COMMANDS – THE COMMAND SHOULD POST A TASK THAT SIGNALS THE EVENT.

Of course, acquiring a high-frequency signal from our example Read implementation is a bit silly. As the implementation is caching a value, sampling it more than once isn't very helpful. But this call pattern – issuing a new request in an event signaling request completion – is a common one.

The problems caused by this signaling raise the question of how FilterMagC is going to signal the readDone event. It needs a way to schedule a function to be called later (like an interrupt). The right way to do this is with a task. This is how our data filter component might look like implemented with a task:

```
module FilterMagC {
  provides interface StdControl;
  provides interface Read<uint16_t>;
  uses interface Timer<TMilli>;
  uses interface Read<uint16_t> as RawRead;
}
implementation {
  uint16_t filterVal = 0;

  ... unchanged ...

  task void readDoneTask() {
    signal Read.readDone(SUCCESS, filterVal);
  }
```

Execution model

```
command error_t Read.read() {
  post readDoneTask();
  return SUCCESS;
}
}
```

Listing 5.4 An improved implementation of FilterMagC

When FilterMagC's Read.read is called, FilterMagC posts readDoneTask and returns immediately. At some point later, TinyOS runs the task, which signals Read.readDone with the filtered value.

5.4 Exercises

1. Measure the CPU cycles it takes to post and run a task. You can do this by using a timer to measure how many times a self posting task can run within a period of time.
2. Sometimes, components introduce a spin loop by posting task that self-posts until a variable changes. Why is this a bad idea, and what's a better solution?
3. Write a singleton loopback interface, where a program can call Send and the component will signal Receive. Be sure to respect buffer-swapping, and be careful of call loops.

6 Applications

You have already seen several of the basic TinyOS components and interfaces, such as booting (MainC, Boot), LED control (LedsC, Leds), and timing (TimerMilliC, Timer). In this chapter, we'll present these components in more detail, along with the other basic TinyOS subsystems (sensing, communication, storage). We introduce and motivate all these subsystems through a running example, a simple anti-theft demo application. By necessity, this anti-theft demo is somewhat platform-specific, as it uses specific light and movement sensors to detect theft. However, as we discuss below, the fact that TinyOS is built over reusable interfaces makes the application trivial to port from the micaz with mts310 sensor board for which it was written to any platform with equivalent communication and sensing capabilities.

Appendix 1 provides a more systematic overview of the major TinyOS services. Each service has a brief description, a list of the components that provide it, and pointers to relevant TinyOS Enhancement Proposals (TEPs) that provide a more detailed specification.

The complete code for the applications in this chapter can be found in TinyOS's contributed code directory (see Section 1.5).

6.1 The basics: timing, LEDs, and booting

One of the simplest anti-theft devices on sale is a simple blinking red light, designed to deter thieves by making them believe that more must be going on.

Such a simple application is, unsurprisingly, trivial to build using TinyOS . We start with the application's main module, AntiTheftC:

```
module AntiTheftC {
  uses {
    interface Boot;
    interface Timer<TMilli> as WarningTimer;
    interface Leds;
  }
}
implementation {
  enum { WARN_INTERVAL = 4096, WARN_DURATION = 64 };
```

```
  event void WarningTimer.fired() {
    if (call Leds.get() & LEDS_LED0)
      { // Red LED is on. Turn it off, will switch on again in 4096-64ms.
        call Leds.led0Off();
        call WarningTimer.startOneShot(WARN_INTERVAL - WARN_DURATION);
      }
    else
      { // Red LED is off. Turn it on for 64ms.
        call Leds.led0On();
        call WarningTimer.startOneShot(WARN_DURATION);
      }
  }

  event void Boot.booted() {
    // We just booted. Perform first LED transition.
    signal WarningTimer.fired();
  }
}
```

Listing 6.1 Anti-theft: simple flashing LED

Our application wants to show that it's active and doing "something" using a red LED. However, an LED is relatively power hungry, so we don't want to leave it on all the time. Instead, we will turn it on for 64 ms every four seconds. We accomplish this by using a single timer: if the LED is off, we switch it on and ask to be woken again in 64 ms; if the LED is on, we switch it off and ask to be woken in 3.936 s. This logic is implemented by the WarningTimer.fired event, based on the commands and events provided by the Timer and Leds interfaces:

```
interface Leds {
  async command void led0On();
  async command void led0Off();
  async command void led0Toggle();

  async command void led1On();
  async command void led1Off();
  async command void led1Toggle();

  async command void led2On();
  async command void led2Off();
  async command void led2Toggle();

  async command uint8_t get();
  async command void set(uint8_t val);
}
```

Listing 6.2 The Leds interface

On a micaz, the red LED has number 0.[1] WarningTimer.fired checks the status of this LED using the Leds.get command, and turns it on or off as appropriate, using the led0On and led0Off commands. After turning the LED on or off, the code also schedules itself to run again after the appropriate interval by using WarningTimer.startOneShot to schedule a one-shot timer.

AntiTheftC needs to start blinking the red LED at boot time, so contains a handler for the booted event from the Boot interface:

```
interface Boot {
    event void booted();
}
```

Listing 6.3 The Boot interface

As we saw earlier, the booted event of the Boot interface (provided by MainC) is signaled as soon as a node has finished booting. All we need to do is execute the logic in WarningTimer.fired to initiate the first LED transition. We could do this by replicating some of the code from WarningTimer.fired in the booted event, but this would be wasteful and, in more complex cases, error-prone. Instead, the simplest approach would be to pretend that the WarningTimer.fired event happened at boot-time. This kind of requirement is not uncommon, so nesC allows modules to signal their own events (and call their own commands), as we see in Boot.booted.

6.1.1 Deadline-based timing

As seen so far, AntiTheftC uses a simple "wake me up in n ms" one-shot timing interface to get periodic events – it cannot use the simpler startPeriodic command as it is interspersing two timing periods (64 ms and 3936 ms). However, such relative-time interfaces (the expiry time is specified as an offset from "now") have a well-known drawback: if one timer event is delayed by some other activity in the system, then all subsequent activities are delayed. If these delays occur frequently, then the system "drifts": instead of blinking the LED every 4 s, it might blink it (on average) every 4.002 s.

This is clearly not a problem for our simple theft-deterrence system, but can be an issue in other contexts. So we show below how the TinyOS timing system allows you to avoid such problems. First, let's see the full Timer interface:

```
interface Timer<precision_tag> {
  // basic interface
  command void startPeriodic(uint32_t dt);
  command void startOneShot(uint32_t dt);
```

---

[1] The LEDs are numbered as LED colors are different on different platforms, e.g. the micaz has red, green, and yellow LEDs while the Telos has red, green, and blue LEDs.

```
        command void stop();
        event void fired();

        // status interface
        command bool isRunning();
        command bool isOneShot();
        command uint32_t gett0();
        command uint32_t getdt();

        // extended interface
        command void startPeriodicAt(uint32_t t0, uint32_t dt);
        command void startOneShotAt(uint32_t t0, uint32_t dt);
        command uint32_t getNow();
    }
```

Listing 6.4 The full Timer interface

We have already seen the commands and events from the basic interface, except for stop which simply cancels any outstanding fired events. The status interface simply returns a timer's current settings. The extended interface is similar to the basic interface, except that timings are specified relative to a base time t0. Furthermore, the "current time" (in the timer's units, since boot) can be obtained using the getNow command. The startOneShotAt command requests a fired event at time t0+dt, and startPeriodicAt requests fired events at times t0+dt, t0+2dt, t0+3dt, ... This may not seem like a very big change, but is sufficient to fix the drift problem as this revised version of WarningTimer.fired shows:

```
uint32_t base; // Time at which WarningTimer.fired should've fired

event void WarningTimer.fired() {
  if (call Leds.get() & LEDS_LED0)
    { // Red LED is on. Turn it off, will switch on again in 4096-64ms.
      call Leds.led0Off();
      call WarningTimer.startOneShotAt(base, WARN_INTERVAL - WARN_DURATION);
      base += WARN_INTERVAL - WARN_DURATION;
    }
  else
    { // Red LED is off. Turn it on for 64ms.
      call Leds.led0On();
      call WarningTimer.startOneShotAt(base, WARN_DURATION);
      base += WARN_DURATION;
    }
}
```

Listing 6.5 WarningTimer.fired with drift problem fixed

The base module variable contains the time at which WarningTimer.fired is expected to fire. By specifying the timer deadline as an offset from base, rather than relative to the current time, we avoid any drift due to WarningTimer.fired running late. We also update base every time we reschedule the timer, and initialize it to the current time in Boot.booted (before starting the first timer):

```
base = call WarningTimer.getNow();
```

The TinyOS timers use 32-bit numbers to represent time. As a result, the millisecond timers (TMilli) wrap around every ∼48.5 days, and a microsecond timer wraps around nearly every hour. These times are shorter than most expected sensor network deployments, so the timers are designed to work correctly when time wraps around. The main user-visible effect of this is that the t0 argument to startOneShotAt and startPeriodicAt is assumed to always represent a time in the past. As a result, values numerically greater than the current time actually represent a time from before the last wrap around.

A final note: on typical sensor node platforms, the millisecond timer is based on a reasonably precise 32 768 Hz crystal. However, this does not mean that timings are perfectly accurate. Different nodes will have slightly different crystal frequencies, which will drift due to temperature effects. If you need time to be really accurate and/or synchronized across nodes, you will need to provide some form of networked time synchronization.

6.1.2 Wiring AntiTheftC

The AntiTheftAppC configuration that wraps AntiTheftC up into a complete application contains no surprises. It instantiates a new timer using the TimerMilliC generic component, and wires AntiTheftC to the LedsC and MainC components that provide LED and booting support:

```
configuration AntiTheftAppC { }
implementation {
  components AntiTheftC, MainC, LedsC;
  components new TimerMilliC() as WTimer;

  AntiTheftC.Boot -> MainC;
  AntiTheftC.Leds -> LedsC;
  AntiTheftC.WarningTimer -> WTimer;
}
```

Listing 6.6 Anti-Theft: application-level configuration

6.2 Sensing

Sensing is, of course, one of the primary activities of a sensor network. However, there are thousands of different sensors out there, measuring everything from light, sound, and acceleration to magnetic fields and humidity. Some sensors are simple analogue sensors, others have digital interfaces or look like a counter (e.g. wind velocity). Furthermore, the precision, sampling rate, and jitter requirements vary greatly from application to application: a simple environmental monitoring application might sample a few sensors every 5 minutes, while a seismic monitoring system might sample acceleration at a few kHz for several seconds, but only once per day.

As a result, TinyOS does not offer a single unified way of accessing all sensors. Instead, it defines a set of common interfaces for sampling sensors, and a set of guidelines for building components that give access to particular sensors.

6.2.1 Simple sampling

The two main sampling interfaces are Read (which we saw in Section 3.4) and ReadStream, covering respectively the case of acquiring single samples and sampling at a fixed rate (with low jitter). As a reminder, Read provides a split-phase read command, that initiates sampling, and a readDone event that reports the sample value and any error that occurred, for an arbitrary type val_t:

```
interface Read<val_t> {
  command error_t read();
  event void readDone(error_t result, val_t val);
}
```

Listing 6.7 The Read interface

In common with most TinyOS split-phase interfaces, you can only start a single sample operation at a time, i.e. calling read again before readDone is signaled will fail (read will not return SUCCESS). However (again in common with other split-phase interfaces), it *is* legal to call read from the readDone event handler, making it easy to perform back-to-back reads.

We can make our anti-theft application more realistic by detecting theft attempts. A first simplistic attempt is based on the observation that stolen items are often placed in bags and pockets, i.e. dark locations. The DarkC module detects dark locations by periodically (every DARK_INTERVAL) sampling a light sensor (using the Read interface) and checking whether the light value is below some threshold (DARK_THRESHOLD). It then reports theft by turning the yellow LED (LED 2 on the micaz) on:

```
module DarkC {
  uses {
    interface Boot;
    interface Leds;
    interface Timer<TMilli> as TheftTimer;
    interface Read<uint16_t> as Light;
  }
}
implementation {
```

```
enum { DARK_INTERVAL = 256, DARK_THRESHOLD = 200 };

event void Boot.booted() {
  call TheftTimer.startPeriodic(DARK_INTERVAL);
}

event void TheftTimer.fired() {
  call Light.read(); // Initiate split-phase light sampling
}

/* Light sample completed. Check if it indicates theft */
event void Light.readDone(error_t ok, uint16_t val) {
  if (ok == SUCCESS && val < DARK_THRESHOLD)
    call Leds.led2On(); /* ALERT! ALERT! */
  else
    call Leds.led2Off(); /* Don't leave LED permanently on */
}
```

Listing 6.8 Anti-theft: detecting dark conditions

We could have implemented dark-detection within the existing AntiTheftC module, and reused the existing WarningTimer. While this would slightly reduce CPU and storage requirements in the timer subsystem, it would increase code complexity in AntiTheftC by mixing code for two essentially unrelated activities. It is generally better to have a single module do one task (blinking an LED, detecting dark conditions) well rather than build a complex module which tries to handle many tasks, paying a significant price in increased complexity.

As is, DarkC is very simple: it initiates periodic sampling in its booted event. Four times a second, TheftTimer.fired requests a new light sample using the split-phase Read interface representing the light sensor (Light). If the sampling succeeds (ok == SUCCESS), then the light is compared to the threshold indicating dark conditions and hence theft. DarkC does not check the error return from Light.read, as there is no useful recovery action when light sampling cannot be initiated – it will retry the detection in 1/4s anyway.

6.2.2 Sensor components

Sensors are represented in TinyOS by generic components offering the Read and/or ReadStream interfaces, and possibly other sensor-specific interfaces (e.g. for calibration). A single component normally represents a single sensor, e.g. PhotoC for the light sensor on the mts310 sensor board:

```
generic configuration PhotoC() {
  provides interface Read<uint16_t>;
}
```

If two sensors are closely related (e.g. the X and Y axis of an accelerometer) they may be offered by a single component. Similarly, if a sensor supports both single (Read) and stream (ReadStream) sampling, the interfaces may be offered by the same generic component. However, neither of these is required: for example, the mts300 sensor board has separate AccelXC, AccelXStreamC, AccelYC and AccelYStreamC components for sampling its two-axis accelerometer. A single component simplifies the namespace, but may lead to extra code and RAM usage in applications that don't need, e.g. both axes or stream sampling.

Adding DarkC to our existing anti-theft application just requires wiring it to the light sensor, and its other services:

```
configuration AntiTheftAppC { }
implementation {
   ... /* the wiring for the blinking red LED */ ...
   components DarkC;
   components new TimerMilliC() as TTimer;
   components new PhotoC();

   DarkC.Boot -> MainC;
   DarkC.Leds -> LedsC;
   DarkC.TheftTimer -> TTimer;
   DarkC.Light -> PhotoC;
}
```

Listing 6.9 Anti-Theft: wiring to light sensor

6.2.3 Sensor values, calibration

TinyOS 2.x specifies the general structure of sensor components such as PhotoC, but, because of the extreme diversity of sensors, does not attempt to specify much else. The type used to report sensor values (uint16_t for PhotoC), the meaning of the values reported by the sensor, the time taken to obtain a sample, the accuracy of sensor values, calibration opportunities and requirements are all left up to the particular sensor hardware and software. Thus, for example, the fact that 200 (DARK_THRESHOLD) is a good value for detecting dark conditions is specific to the particular photo-resistor used on the mts300 board, and to the way it is connected to micaz motes (in series with a specific resistor, connected to the micaz's microcontroller's A/D converter).

In some cases, e.g. temperature, it would be fairly easy to specify a standard interface, such as temperature in $1/10\,K$. However, forcing such an interface on temperature sensors might not always be appropriate: the sensor might be more precise than $1/10\,K$, or the code for doing the conversion to these units might take too much time or space on the mote, when the conversion could be done just as easily when the data is recovered from the sensor network. TinyOS leaves these decisions to individual sensor component designers.

6.2.4 Stream sampling

The ReadStream interface is more complex than Read, in part because it needs to support motes with limited RAM: some applications need to sample more data at once than actually fits in memory. However, simple uses of ReadStream remain quite simple, as we will see by building an alternate theft-detection mechanism to DarkC. First, let's see ReadStream:

```
interface ReadStream<val_t> {
    command error_t postBuffer(val_t* buf, uint16_t count);
    command error_t read(uint32_t usPeriod);
    event void bufferDone(error_t result, val_t* buf, uint16_t count);
    event void readDone(error_t result, uint32_t usActualPeriod);
}
```

Listing 6.10 ReadStream Interface

Like Read, ReadStream is a typed interface whose val_t parameter specifies the type of individual samples. Before sampling starts, one or more buffers (arrays of val_t) must be posted with the postBuffer command. Sampling starts once the read command is called (usPeriod is the sampling period in microseconds), and continues until all posted buffers are full or an error occurs, at which point readDone is signaled. It is also possible to post new buffers during sampling – this is often done in the bufferDone event which is signaled as each buffer is filled up. By using two (or more) buffers, and processing (e.g. computing statistics, or writing to flash) and reposting each buffer in bufferDone, a mote can sample continuously more data than can fit in RAM.

The simplest way to use ReadStream is to declare an array holding N sample values, post the buffer and call read. The samples are available once readDone is signaled. We use this approach in MovingC, an alternate component to DarkC that detects using an accelerometer to detect movement – it's hard to steal something without moving it. MovingC samples acceleration at 100 Hz for 1/10 s, and reports theft when the variance of the sample is above a small threshold (picked experimentally):

```
module MovingC {
  uses {
    interface Boot;
    interface Leds;
    interface Timer<TMilli> as TheftTimer;
    interface ReadStream<uint16_t> as Accel;
  }
}
implementation {
  enum { ACCEL_INTERVAL = 256,   /* Checking interval */
         ACCEL_PERIOD = 10000,   /* uS -> 100Hz */
         ACCEL_NSAMPLES = 10,    /* 10 samples * 100Hz -> 0.1s */
         ACCEL_VARIANCE = 4 };   /* Determined experimentally */
```

Applications

```
uint16_t accelSamples[ACCEL_NSAMPLES];
task void checkAcceleration();

event void Boot.booted() {
  call TheftTimer.startPeriodic(ACCEL_INTERVAL);
}

event void TheftTimer.fired() {
  // Get 10 samples at 100Hz
  call Accel.postBuffer(accelSamples, ACCEL_NSAMPLES);
  call Accel.read(ACCEL_INTERVAL);
}

/* The acceleration read completed. Post the task to check for theft */
event void Accel.readDone(error_t ok, uint32_t usActualPeriod) {
  if (ok == SUCCESS)
    post checkAcceleration();
}

/* Check if acceleration variance above threshold */
task void checkAcceleration() {
  uint8_t i;
  uint32_t avg, variance;

  for (avg = 0, i = 0; i < ACCEL_NSAMPLES; i++) avg += accelSamples[i];
  avg /= ACCEL_NSAMPLES;
  for (variance = 0, i = 0; i < ACCEL_NSAMPLES; i++)
    variance+=(int16_t)(accelSamples[i]-avg)*(int16_t)(accelSamples[i]-avg);

  if (variance > ACCEL_VARIANCE * ACCEL_NSAMPLES)
    call Leds.led2On(); /* ALERT! ALERT! */
  else
    call Leds.led2Off(); /* Don't leave LED permanently on */
}

event void Accel.bufferDone(error_t ok, uint16_t *buf, uint16_t count) { }
}
```

Listing 6.11 Anti-theft: detecting movement

The basic structure of MovingC is identical to DarkC: sampling is initiated four times a second (TheftTimer.fired) and, if sampling is successful, the samples are checked to see if they indicate theft (Accel.readDone). The three main differences are:

- postBuffer is called to register the acceleration buffer before sampling starts
- MovingC must implement the bufferDone event, which is signaled when each posted buffer is full – here we do not need to do anything as we are sampling into a single buffer

- the samples are checked in a separate task (checkAcceleration), to avoid making the execution of Accel.readDone take too long (Section 5.2.2)

The wiring for MovingC is identical to DarkC, except that it wires to a streaming accelerometer sensor:

```
components new AccelXStreamC();
...
MovingC.Accel -> AccelXStreamC;
```

6.3 Single-hop networking

TinyOS uses a typical layered network structure, as shown in Figure 6.1. The networking stack is composed of a set of layers, each of which defines a header and footer layout, and includes a variable-size payload space for the layer above; the highest layer (usually the application) just holds the application data. Each layer's header typically includes a "type" field that specifies the meaning of that layer's payload and hence the payload's layout – this makes it easy to use multiple independent packets in a single application. For instance, in Figure 6.1 there are two kinds of packets: packets containing messages for application 1, built over layer2a, itself built over layer1 and packets containing messages for application 3, built over layer2b, itself built over layer1.

The lowest networking layer exposed in TinyOS is called *active messages* (AM) [29]. AM is typically implemented directly over a mote's radio, and provides unreliable, single-hop packet transmission and reception. Each packet is identified by an *AM type*, an 8-bit integer that identifies the packet type. The name "active messages" comes from the fact that the type is used to automatically dispatch received packets to an appropriate handler: in a sense, packets (messages) are active because they identify (via their type) code to be executed.

A variable of type message_t holds a single AM packet. Each packet can hold a user-specified *payload* of up to TOSH_DATA_LENGTH bytes – normally 28 bytes,

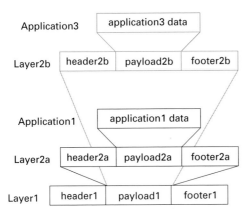

Figure 6.1 TinyOS packet layout.

but this can be changed at compile-time up to 255 bytes.[2] Note that increasing TOSH_DATA_LENGTH increases the size of every message_t variable so may cause substantial RAM usage increases.

Some simple mote applications are built directly over AM, whilst more complex applications typically use higher-level protocols such as dissemination and collection which we discuss in the next section. These higher-level protocols are themselves built over AM. It is worth noting that the interfaces used to access AM (AMSend, Receive) are sometimes reused in some higher-level protocols (e.g. tree collection uses Receive) – this makes switching between protocols easier and encourages code reuse.

6.3.1 Sending packets

Our anti-theft application currently reports theft by lighting a LED, which is unlikely to be very effective. A much better approach would be to alert someone, e.g. by sending them a radio message. This message should contain as its payload the identity of the node being stolen. As discussed in Section 3.5.3, TinyOS payloads are always defined using platform-independent types to ensure interoperability. So the definition of the theft report payload is a simple platform-independent struct, declared in the `antitheft.h` header file:

```
#ifndef ANTITHEFT_H
#define ANTITHEFT_H

typedef nx_struct theft {
  nx_uint16_t who;
} theft_t;
...
#endif
```

The AMSend ("active message send") interface contains all the commands needed to fill-in and send packets:

```
interface AMSend {
  command error_t send(am_addr_t addr, message_t* msg, uint8_t len);
  event void sendDone(message_t* msg, error_t error);
  command error_t cancel(message_t* msg);

  command uint8_t maxPayloadLength();
  command void* getPayload(message_t* msg, uint8_t len);
}
```

Listing 6.12 The AMSend interface

---

[2] Some radios may impose a lower limit.

The send command and sendDone event are used to perform split-phase transmission of an AM packet stored in a message_t packet buffer; the addr parameter specifies the packet destination and may be TOS_BCAST_ADDR to request a broadcast. AM packets have source addresses, which can be retrieved with the AMPacket interface. A node's AM address is usually – but not always – the same as its TOS_NODE_ID. Technically, TOS_NODE_ID is a unique identifier for a node, while the AM address is the address of its radio interface. A node with two radios, for example, could have different AM addresses on the two.

Because the user-payload of a packet buffer may be at different offsets on different platforms (with different radios) or when using different communication layers, AMSend provides a getPayload command to obtain a pointer to the payload. The getPayload command requires that the user pass the size of their payload so that it can ensure that there is enough space in the buffer (it returns NULL if there isn't enough space) – applications wishing to send variable-sized messages can call maxPayloadLength to find the maximum possible payload size.

Putting all this together, the reportTheft function sends theft-report packets to any nearby listeners:

```
uses interface AMSend as Theft;
...
message_t reportMsg;  // The theft report message buffer
bool sending;         // Don't try and send while a send is in progress

void reportTheft() {
  theft_t *payload = call Theft.getPayload(&reportMsg, sizeof(theft_t));

  if (payload && !sending)
    { // We can send if we're idle and the payload fits
      payload->who = TOS_NODE_ID; // Report that *we* are being stolen!
      // And send the report to everyone (TOS_BCAST_ADDR)
      if(call Theft.send(TOS_BCAST_ADDR, &reportMsg, sizeof(theft_t))==SUCCESS)
        sending = TRUE;
    }
}

event void Theft.sendDone(message_t *msg, error_t error) {
  sending = FALSE; // Our send completed
}
```

Listing 6.13 Anti-Theft: reporting theft over the radio

The reportTheft function refuses to attempt a send if its payload doesn't fit in reportMsg (payload == NULL), or if a previous send is still in progress (sending is true): during transmission the packet buffer is "owned" (Section 3.5.1) by the communication stack, so should not be touched. The code checks for failure in transmission (so that it can update sending correctly), but doesn't otherwise attempt to retry failed transmissions, or to check whether the theft report was received by anyone. It simply relies on the fact that MovingC will call reportTheft repeatedly when a "theft" is in progress, so one of the messages is likely to get through:

```
if (variance > ACCEL_VARIANCE * ACCEL_NSAMPLES)
{
   call Leds.led2On();  /* ALERT! ALERT! */
   reportTheft();
}
```

Most high-level TinyOS subsystems are started and stopped automatically, either when the system boots, or on-demand based on usage. This is however not true of the AM-based radio and serial communication stacks for two main reasons. First, communication is typically power-hungry (e.g. most radios use as much power as an active microcontroller even when not actively receiving or transmitting packets), so leaving a communication stack permanently on is not necessarily a good idea. Second, AM-based packet reception is inherently asynchronous: unlike, e.g. sampling which is application-driven, there is nothing in the code on a mote that specifies when packets are expected to be received. Thus, the communication stack cannot switch itself on and off on-demand. Instead, applications must use the split-phase SplitControl interface we saw earlier to start and stop the radio:

```
interface SplitControl {
  command error_t start();
  event void startDone(error_t error);

  command error_t stop();
  event void stopDone(error_t error);
}
```

Listing 6.14 The SplitControl interface

MovingC switches the radio on at boot-time, and waits until the radio starts before initiating sampling to avoid sending a message while the radio is off:

```
uses interface SplitControl as CommControl;
...
event void Boot.booted() {
  call CommControl.start();
}

event void CommControl.startDone(error_t ok) {
  // Start checks once communication stack is ready
  call TheftTimer.startPeriodic(ACCEL_INTERVAL);
}

event void CommControl.stopDone(error_t ok) { }
```

This code ignores the error code in startDone, as there's not any obvious recovery step to take if the radio will not start.

6.3.2 Receiving packets

MovingC uses hard-wired constants for the check intervals and the acceleration-variance threshold. If these are inappropriate for some contexts (e.g. a moving vehicle may have some background vibration), then the code has to be changed and recompiled with new values, and then reinstalled in the sensor network. A better approach is to change the code once to allow remote configuration, using AM packets with the following payload (also declared in `antitheft.h`):

```
typedef nx_struct settings {
  nx_uint16_t accelVariance;
  nx_uint16_t accelInterval;
} settings_t;
```

AM Packets reception is provided by the TinyOS Receive interface (also seen earlier):

```
interface Receive {
  event message_t* receive(message_t* msg, void* payload, uint8_t len);
}
```

Listing 6.15 The Receive interface

As discussed in Section 3.5.1, implementations of Receive.receive receive a packet buffer which they can either simply return, or hang onto as long as they return a different buffer. In the case of MovingC we choose the first option, as we are done with the settings once we have read them out of the packet:

```
uses interface Receive as Settings;
...
uint16_t accelVariance = ACCEL_VARIANCE;

event message_t *Settings.receive(message_t *msg,
    void *payload, uint8_t len) {
  if (len >= sizeof(settings_t)) // Check the packet seems valid
    { // Read settings by casting payload to settings_t, reset check interval
      settings_t *settings = payload;

      accelVariance = settings->accelVariance;
      call TheftTimer.startPeriodic(settings->accelInterval);
    }
  return msg;
}

task void checkAcceleration() {
  ...
  if (variance > accelVariance * ACCEL_NSAMPLES)
    {
```

Applications

```
        call Leds.led2On(); /* ALERT! ALERT! */
        reportTheft();
    }
}
```

Listing 6.16 Anti-Theft: changing settings

6.3.3 Selecting a communication stack

Nothing in the code seen so far specifies which communication stack is used for theft reports, or which AM types distinguish theft reports from settings updates. This information is specified in configurations, by wiring to the components representing the desired communication stack: these components have compatible signatures, making it easy to switch between stacks:

```
configuration ActiveMessageC {                    configuration SerialActiveMessageC{
  provides interface SplitControl;                  provides interface SplitControl;
  ...                                               ...
}                                                 }
generic configuration AMSenderC(am_id_t id) {     generic configuration SerialAMSenderC(...){
  provides interface AMSend;                        provides interface AMSend;
  ...                                               ...
}                                                 }
generic configuration AMReceiverC(am_id_t id) {   generic configuration SerialAMReceiverC(...){
  provides interface Receive;                       provides interface Receive;
  ...                                               ...
}                                                 }
```

Listing 6.17 Serial vs Radio-based AM components

ActiveMessageC contains the SplitControl interface to start and stop the communication stack. Instantiating AMSenderC with a particular AM type creates a component that provides an AMSend interface that will queue a single packet for transmission, independently of all other AMSenderC instances. This makes life simpler for modules using AMSend: as long as they only send one packet at a time, they can be assured that they can always enqueue their packet. Finally, instantiating AMReceiverC with AM type k and wiring the Receive interface to your code automatically dispatches received packets of type k to your Receive.receive event handler.

The anti-theft application uses radio-based communication (the mts300 sensor board physically interferes with serial connections):

```
    enum { AM_THEFT = 42, AM_SETTINGS = 43 };
    ...
    components ActiveMessageC;
    components new AMSenderC(AM_THEFT) as SendTheft;
    components new AMReceiverC(AM_SETTINGS) as ReceiveSettings;
    MovingC.CommControl -> ActiveMessageC;
```

```
MovingC.Theft -> SendTheft;
MovingC.Settings -> ReceiveSettings;
```

6.4 Multi-hop networking: collection, dissemination, and base stations

A mote network is typically more complex than a set of motes within direct radio range (often only 30m/100ft on current motes) of each other. Instead, *multi-hop* networks use motes to relay messages when the sender and receiver are not in direct radio range. Furthermore, one or more *base station* motes physically connected to a PC-class device, and usually line-powered, relay information to and from the wider world. Thus, considered as a whole, a sensor network application has three parts (Figure 6.2):

- Mote code: the code running on the motes in the network. Interactions between motes take the form of sending and receiving radio messages.
- Base Station code: the code running on the base station mote. It interacts with the other motes via radio messages, and exchanges packets with the PC over a serial connection.
- PC code: the code running on the PC.

TinyOS provides two basic multi-hop networking abstractions: tree collection, and dissemination. In tree collection, the motes organize themselves into a routing tree centered on a particular mote, the *root*, which is often a base station mote. All messages sent in the tree automatically flow to the root (e.g. following the links shown in Figure 6.2). Collection trees are typically used to collect information (e.g. sensor data) from a sensor network.

Dissemination efficiently distributes a value (which can be a structure with several fields) across the whole mote network. Furthermore, any mote can update the value, and the whole network will eventually settle on the value from the most recent update. Dissemination is often used for run-time configuration of mote applications.

In the rest of this section, we'll adapt AntiTheft to use collection and dissemination rather than ActiveMessageC for its theft report and configuration. In particular, we show

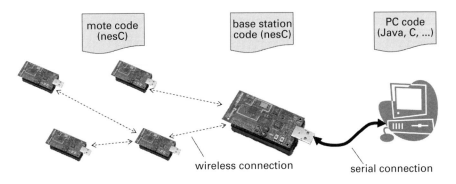

Figure 6.2 A typical sensor network.

how the base station mote sets itself up as the root of the collection tree to report theft settings to the PC, and how it disseminates new settings received from the PC. Chapter 7 presents the PC-side of the mote connection.

6.4.1 Collection

Sending a message via a collection tree is very similar to using AM, except that messages do not have a destination address (the tree root is the implicit destination). Thus collection trees use the Send interface, which is identical to AMSend except for the lack of an addr parameter to the send command:

```
interface Send {
  command error_t send(message_t* msg, uint8_t len);
  event void sendDone(message_t* msg, error_t error);
  command error_t cancel(message_t* msg);

  command uint8_t maxPayloadLength();
  command void* getPayload(message_t* msg, uint8_t len);
}
```

Listing 6.18 The Send interface

As a result, the code in MovingC for reporting a theft over a collection tree is nearly identical to that from Section 6.3.1:

```
uses interface Send as Theft;
...
message_t reportMsg;  // The theft report message buffer
bool sending;         // Don't try and send while a send is in progress

void reportTheft() {
  theft_t *payload = call Theft.getPayload(&reportMsg, sizeof(theft_t));

  if (payload && !sending)
    { // We can send if we're idle and the payload fits
      payload->who = TOS_NODE_ID; // Report that *we* are being stolen!
      // And send the report to the root
      if (call Theft.send(&reportMsg, sizeof(theft_t)) == SUCCESS)
        sending = TRUE;
    }
}

event void Theft.sendDone(message_t *msg, error_t error) {
  sending = FALSE; // Our send completed
}
```

Listing 6.19 Anti-Theft: reporting theft over a collection tree

6.4.2 Dissemination

The dissemination service is accessed by an interface parameterized by the type of value being disseminated:

```
interface DisseminationValue<t> {
  command const t* get();
  command void set( const t* );
  event void changed();
}
```

Listing 6.20 DisseminationValue interface

The dissemination values are sent across the network, so the type passed to DisseminationValue should be a platform-independent type (Section 3.5.3). In our case, we reuse the settings_t type we defined in Section 6.3.2, which contains new acceleration variance and check intervals. The code in MovingC to receive and handle new settings is simple:

```
uses interface DisseminationValue<settings_t> as Settings;
...
/* New settings received, update our local copy */
event void Settings.changed() {
  const settings_t *newSettings = call Settings.get();

  accelVariance = newSettings->accelVariance;
  call TheftTimer.startPeriodic(newSettings->accelInterval);
}
```

Listing 6.21 Anti-Theft: settings via a dissemination tree

6.4.3 Wiring collection and dissemination

Like AM itself, the collection and dissemination services must be explicitly started, this time via the non-split-phase StdControl interface:

```
interface StdControl {
  command error_t start();
  command error_t stop();
}
```

Listing 6.22 The StdControl interface

Furthermore, as they are built over AM, the radio must be started first. As a result, the AntiTheft boot sequence is now slightly more complex (as earlier, we don't check error codes as there is no obvious recovery step):

```
uses interface SplitControl as CommControl;
uses interface StdControl as CollectionControl;
uses interface StdControl as DisseminationControl;
...
event void Boot.booted() {
  call CommControl.start();
}

event void CommControl.startDone(error_t ok) {
  // Start multi-hop routing and dissemination
  call CollectionControl.start();
  call DisseminationControl.start();
  // Start checks once communication stack is ready
  call TheftTimer.startPeriodic(ACCEL_INTERVAL);
}
```

AM types identify different kinds of messages in a mote network. In a similar fashion, collection and dissemination use 8-bit identifiers to allow for multiple independent collection trees and for multiple values to be disseminated. As with AM, these identifiers are specified as arguments to the CollectionSenderC and DisseminatorC generic components that provide access to collection and dissemination:

```
enum { COL_THEFT = 54, DIS_THEFT = 55 };
...
components ActiveMessageC, DisseminationC, CollectionC;
MovingC.CommControl -> ActiveMessageC;
MovingC.CollectionControl -> CollectionC;
MovingC.DisseminationControl -> DisseminationC;

/* Instantiate and wire our collection service for theft alerts */
components new CollectionSenderC(COL_THEFT) as TheftSender;
MovingC.Theft -> TheftSender;

/* Instantiate and wire our dissemination service for theft settings */
components new DisseminatorC(settings_t, DIS_THEFT);
MovingC.Settings -> DisseminatorC;
```

6.4.4 Base station for collection and dissemination

For collection and dissemination to be useful, something must consume the messages sent up the collection tree, and produce new values to disseminate. In the case of AntiTheft, we assume a PC with a serial connection to a base station mote displays theft reports and allows the theft detection settings to be changed. Here we just show how the base station mote is setup to forward tree collection theft reports and disseminate theft settings; Chapter 7 presents the libraries and utilities that are used to write PC applications that communicate with mote networks.

6.4 Networking: collection, dissemination

The base station for AntiTheft is a separate nesC program. In other applications, base stations and regular motes run the same nesC code but are distinguished using some other means (e.g. many applications specify that a mote with identifier 0 is a base station).

A base station mote communicates with the PC via AM over the serial port, using SerialActiveMessageC. The mote and the PC exchange settings (PCSettings) and theft reports (PCTheft):

```
configuration AntiTheftRootAppC { }
implementation {
  components AntiTheftRootC;

  components SerialActiveMessageC,
    new SerialAMReceiverC(AM_SETTINGS) as PCSettings,
    new SerialAMSenderC(AM_THEFT) as PCTheft;

  AntiTheftRootC.SerialControl -> SerialActiveMessageC;
  AntiTheftRootC.RSettings -> PCSettings;
  AntiTheftRootC.STheft -> PCTheft;
  ...
}
```

When the base station receives new settings, it simply calls the change command in the DisseminationUpdate interface:

```
interface DisseminationUpdate<t> {
  command void change(t* newVal);
}
```

Listing 6.23 The DisseminationUpdate interface

The call to change automatically triggers the dissemination process:

```
module AntiTheftRootC {
 uses interface DisseminationUpdate<settings_t> as USettings;
 uses interface Receive as RSettings;
 ...
 /* When we receive new settings from the serial port, we disseminate
    them by calling the change command */
 event message_t *RSettings.receive(message_t* msg, void* payload, uint8_t len)
 {
   if (len == sizeof(settings_t))
     call USettings.change((settings_t *)payload);
   return msg;
 }
}
```

Listing 6.24 AntiTheft base station code: disseminating settings

To receive theft reports, the base station mote must mark itself as the root of the collection tree using the RootControl interface:

```
interface RootControl {
    command error_t setRoot();
    command error_t unsetRoot();
    command bool isRoot();
}
```

Listing 6.25 The RootControl interface

Once a mote is the root of the collection tree, it receives the messages sent up the tree via a regular Receive interface. In the case of the AntiTheft base station mote, it simply forwards the payload of these messages (a theft_t value) to the serial port via its STheft interface:

```
module AntiTheftRootC {
  uses interface RootControl;
  uses interface Receive as RTheft;
  uses interface AMSend as STheft;
  ...

  event void CommControl.startDone(error_t error) {
    ...
    // Set ourselves as the root of the collection tree
    call RootControl.setRoot();
  }

  message_t fwdMsg;
  bool fwdBusy;

  /* When we (as root of the collection tree) receive a new theft alert,
     we forward it to the PC via the serial port */
  event message_t *RTheft.receive(message_t* msg, void* payload, uint8_t len)
  {
    if (len == sizeof(theft_t) && !fwdBusy)
      {
        /* Copy payload from collection system to our serial message buffer
           (fwdTheft), then send our serial message */
        theft_t *fwdTheft = call STheft.getPayload(&fwdMsg, sizeof(theft_t));
        if (fwdTheft != NULL) {
          *fwdTheft = *(theft_t *)payload;
          if(call STheft.send(TOS_BCAST_ADDR, &fwdMsg, sizeof*fwdTheft)==SUCCESS)
            fwdBusy = TRUE;
        }
      }
    return msg;
  }
}
```

```
    fwdBusy = FALSE;
  }
}
```

Listing 6.26 AntiTheft base station code: reporting thefts

The base station application must wire the serial port (already shown), and the collection and dissemination interfaces used in the code above. Reception of messages from a tree is specified by indexing a *parameterized interface* with the tree collection identifier; parameterized interfaces are presented in detail in Section 8.3. The resulting collection and dissemination wiring is:

```
configuration AntiTheftRootAppC { }
implementation {
  ... boot, serial and radio wiring ...

  components DisseminationC, new DisseminatorC(settings_t, DIS_THEFT);
  AntiTheftRootC.DisseminationControl -> DisseminationC;
  AntiTheftRootC.USettings -> DisseminatorC;

  components CollectionC;
  AntiTheftRootC.CollectionControl -> CollectionC;
  AntiTheftRootC.RootControl -> CollectionC;
  AntiTheftRootC.RTheft -> CollectionC.Receive[COL_THEFT];
}
```

Listing 6.27 AntiTheft base station wiring

6.5 Storage

Many motes include some amount of flash-based non-volatile storage, e.g. 512 kB on the micaz. TinyOS divides this non-volatile storage into volumes. A volume is a contiguous region of storage with a certain format that can be accessed with an associated interface. TinyOS defines three basic storage abstractions: Log, Block, and Config. Log is for append-only writes and streaming reads, Block is for random-access reads and write, and Config is for small items of configuration data. Log and Config have the advantage that their more limited interface allows for atomic operations: when a write to a Log or Config volume completes, it is guaranteed to be written. In contrast, the Block interface has a separate commit operation.

TinyOS uses this abstraction-on-a-volume approach rather than a more traditional filing system for two main reasons. First, it is simpler, reducing code and RAM requirements for applications using permanent storage. Second, sensor networks are normally dedicated to a single application, which should have a reasonable knowledge

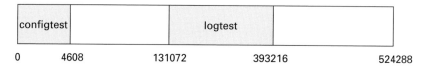

Figure 6.3 Sample volume specification and resulting flash layout for a micaz mote.

of its storage requirements. Thus, the full generality of a filing system is typically not required.

6.5.1 Volumes

The division of a mote's flash chip into volumes is specified at compile-time, by a simple XML configuration file. Flash chips have different sizes, and different rules on how they can be divided, so the volume specification is necessarily chip-specific. By convention, the volume configuration for chip C is found in a file named volumes-C.xml. For instance, Figure 6.3 shows a volumes-at45db.xml file specifying two volumes for the Atmel AT45DB chip found on the micaz (see TEP 103 [4] for a specification of the volume configuration format). The first volume is named LOGTEST, starts at an offset of 128kB and is 256 kB long. The second is named CONFIGTEST and is 4.5kB with no offset specified: some offset will be picked at compile-time.

A storage abstraction is mapped to a specific volume by instantiating the component implementing the abstraction (LogStorageC, BlockStorageC, and ConfigStorageC) with the volume identifier (prefixed with VOLUME_) as argument, e.g.:

```
generic configuration ConfigStorageC(volume_id_t volid) {
  provides interface Mount;
  provides interface ConfigStorage;
} ...

components new ConfigStorageC(VOLUME_CONFIGTEST) as MyConfiguration;
```

Listing 6.28 ConfigStorageC signature

A volume can be associated with at most one instance of one storage abstraction. A storage abstraction instance has to be associated with a volume so that the underlying code can generate an absolute offset into the chip from a relative offset within a volume. For instance, address 16k on volume LOGTEST is address 144 k on the AT45DB.

6.5.2 Configuration data

As it currently stands, anti-theft motes lose their settings when they are switched off. We can fix this by storing the settings in a small configuration volume:

```
<volume_table>
  <volume name="AT_SETTINGS" size="512"/>
</volume_table>

configuration AntiTheftAppC { }
implementation {
  ...
  components new ConfigStorageC(VOLUME_AT_SETTINGS) as AtSettings;
  MovingC.Mount -> AtSettings;
  MovingC.ConfigStorage -> AtSettings;
}
```

Configuration volumes must be mounted before use, using the split-phase Mount interface:

```
interface Mount {
  command error_t mount();
  event void mountDone(error_t error);
}
```

Listing 6.29 Mount interface for storage volumes

Once mounted, the volume is accessed using the ConfigStorage interface:

```
interface ConfigStorage {
 command error_t read(storage_addr_t addr, void* buf, storage_len_t len);
 event void readDone(storage_addr_t addr, void* buf, storage_len_t len,
          error_t error);

 command error_t write(storage_addr_t addr, void* buf, storage_len_t len);
 event void writeDone(storage_addr_t addr, void* buf, storage_len_t len,
          error_t error);
 command error_t commit();
 event void commitDone(error_t error);

 command storage_len_t getSize();
 command bool valid();
}
```

Listing 6.30 ConfigStorage interface

ConfigStorage volumes are somewhat unusual in that they effectively keep two copies of their contents: the contents as of the last commit and the new, as-yet-to-be-committed

104 Applications

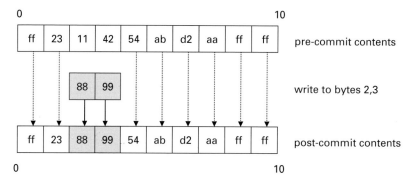

Figure 6.4 ConfigStorage: writes and commits.

contents. The purpose here is to guarantee that configuration data is not lost: if a mote crashes during configuration updates, the data last committed is still present and uncorrupted. Reads always read the last-committed data, while writes update the new copy – you cannot see the data that has been written until after a successful commit. However, writes are still updates: if a configuration volume contains 10 bytes, and bytes 2 and 3 are updated with a write and the volume is then committed, the new contents are bytes 0–1 and 4–9 from before the write, and bytes 2–3 from the write, as shown in Figure 6.4. Because configuration volumes must store multiple copies of data to provide reliability, the actual available size is less than the space reserved for them in the volumes file. The actually available size is returned by the getSize command.

While configuration volumes aim to prevent data loss, there are still two cases where they may not contain valid data: when the volume is first created,[3] or if the flash contents get corrupted by some external physical process (hopefully unlikely). The valid command can be used to find out whether a configuration volume contains valid data.

Putting all this together, adapting MovingC to have permanent settings is not too hard: the configuration volume is mounted and read at boot time, and new settings are saved when they are received. The following simplified excerpts show the basic boot-time logic, with error-checking removed for simplicity:

```
uses interface Mount;
uses interface ConfigStorage;
...
settings_t settings;

event void Boot.booted() {
  settings.accelVariance = ACCEL_VARIANCE; // default settings
  settings.accelInterval = ACCEL_INTERVAL;
  call Mount.mount();
}
```

---

[3] This really means: when this division of the flash is first used on this particular mote – there is no actual "format"-like operation.

```
event void Mount.mountDone(error_t ok) {
  if (call ConfigStorage.valid())
    call ConfigStorage.read(0, &settings, sizeof settings);
  else
    call CommControl.start();
}

event void ConfigStorage.readDone(storage_addr_t addr, void* buf,
                                  storage_len_t len, error_t error) {
  call CommControl.start();
}
```

Listing 6.31 Anti-Theft: reading settings at boot time

The current settings are simply saved as a module-level settings_t variable, which is read if the volume is valid and left with default values if not. Updating the settings simply involves calling write and commit (again with error-checking removed):

```
event void Settings.changed() {
  settings = *call Settings.get();
  call ConfigStorage.write(0, &settings, sizeof settings);
}

event void ConfigStorage.writeDone(storage_addr_t addr, void* buf,
                                   storage_len_t len, error_t error){
  call ConfigStorage.commit();
}

event void ConfigStorage.commitDone(error_t error) {
  call TheftTimer.startPeriodic(settings.accelInterval);
}
```

Listing 6.32 Anti-Theft: saving configuration data

6.5.3 Block and Log storage

Block and Log storage are two abstractions for storing large amounts of data. Log is intended for logging: it provides reliability (each write is a separate transaction), at the cost of limited random access: writes are append-only, reads can only seek to recorded positions. Block is a lower-level abstraction which allows random reads and writes, but provides no reliability guarantees. Also, Block only allows any given byte to be written once between two whole-volume erases. Block is often used to store large items, such as programs for Deluge, the network reprogramming system.

In this section, we will use Block and Log to build FlashSampler, a two-level sampling application. The system will periodically collect 32kB samples of an accelerometer, storing the data in a Block storage volume. It will then log a summary of this sample to a circular log. As a result, the system will at all times have the most recent measurement with full fidelity, and some number of older measurements with reduced fidelity.

For the micaz with the same mts300 sensor board used for the anti-theft application, the following volumes file specifies FlashSampler's log and block volumes – 64 kB for the latest sample (SAMPLES) and the rest for the long-term log (SAMPLELOG):

```
<volume_table>
  <volume name="SAMPLES" size="65536"/>
  <volume name="SAMPLELOG" size="458752"/>
</volume_table>
```

Like the other storage abstractions, block storage is accessed by instantiating the BlockStorageC component with the volume identifier and using its BlockRead and BlockWrite interfaces:

```
generic configuration BlockStorageC(volume_id_t volid) {
  provides interface BlockWrite;
  provides interface BlockRead;
}
```

Listing 6.33 BlockStorageC signature

To sample to flash, we use the the BlockWrite interface:

```
interface BlockWrite {
  command error_t write(storage_addr_t addr, void* buf, storage_len_t len);
  event void writeDone(storage_addr_t addr, void* buf, storage_len_t len,
              error_t error);

  command error_t erase();
  event void eraseDone(error_t error);

  command error_t sync();
  event void syncDone(error_t error);
}
```

Listing 6.34 The BlockWrite interface

All commands are split-phase: erase erases the flash before the first use, write writes some bytes, and sync ensures that all outstanding writes are physically present on the flash.

FlashSampler uses BlockWrite and the ReadStream sampling interface we saw earlier to simultaneously sample and save results to the flash. For this to work, it needs to use two buffers: while one is being sampled, the other is written to the flash. The

6.5 Storage

maximum sampling rate will thus be limited by both the sensor's maximum sampling rate and the flash's maximum write rate. This scheme is implemented in the following AccelSamplerC component, as a split-phase sample-to-flash command provided by a Sample interface (not shown):

```
// in flashsampler.h:
enum {
  SAMPLE_PERIOD = 1000,
  BUFFER_SIZE = 512,     // samples per buffer
  TOTAL_SAMPLES = 32768, // must be multiple of BUFFER_SIZE
};

module AccelSamplerC
{
  provides interface Sample;
  uses interface ReadStream<uint16_t> as Accel;
  uses interface BlockWrite;
}
implementation
{
  uint16_t buffer1[BUFFER_SIZE], buffer2[BUFFER_SIZE];
  int8_t nbuffers; // how many buffers have been filled

  command void Sample.sample() {
    // Sampling requested, start by erasing the block
    call BlockWrite.erase();
  }

  event void BlockWrite.eraseDone(error_t ok) {
    // Block erased. Post both buffers and initiate sampling
    call Accel.postBuffer(buffer1, BUFFER_SIZE);
    call Accel.postBuffer(buffer2, BUFFER_SIZE);
    nbuffers = 0;
    call Accel.read(SAMPLE_PERIOD);
  }

  event void Accel.bufferDone(error_t ok, uint16_t *buf, uint16_t count) {
    // A buffer is full. Write it to the block
    call BlockWrite.write(nbuffers * sizeof buffer1, buf, sizeof buffer1);
  }

  event void BlockWrite.writeDone(storage_addr_t addr, void* buf,
                                  storage_len_t len, error_t error) {
    // Buffer written. TOTAL_SAMPLES is a multiple of BUFFER_SIZE, so
    // once we've posted TOTAL_SAMPLES / BUFFER_SIZE buffers we're done.
    // As we started by posting two buffers, the test below includes a -2
    if (++nbuffers <= TOTAL_SAMPLES / BUFFER_SIZE - 2)
      call Accel.postBuffer(buf, BUFFER_SIZE);
    else if (nbuffers == TOTAL_SAMPLES / BUFFER_SIZE)
      // Once we've written all the buffers, flush writes to the buffer
      call BlockWrite.sync();
  }
```

```
event void BlockWrite.syncDone(error_t error) {
  signal Sample.sampled(error);
}

event void Accel.readDone(error_t ok, uint32_t usActualPeriod) {
  // If we didn't use all buffers something went wrong, e.g., flash writes were
  // too slow, so the buffers did not get reposted in time
  signal Sample.sampled(FAIL);
}
}
```

Listing 6.35 Simultaneously sampling and storing to flash (most error checking omitted)

The second part of FlashSampler reads a sampled block, summarizes it, and writes it to a circular log. The summary is simply a 128× downsample (by averaging) of the original sample, and is thus 512 bytes long.

Reading the block is done using the BlockRead interface:

```
interface BlockRead {
  command error_t read(storage_addr_t addr, void* buf, storage_len_t len);
  event void readDone(storage_addr_t addr, void* buf, storage_len_t len,
            error_t error);

  command error_t computeCrc(storage_addr_t addr, storage_len_t len,
            uint16_t crc);
  event void computeCrcDone(storage_addr_t addr, storage_len_t len,
            uint16_t crc, error_t error);

  command storage_len_t getSize();
}
```

Listing 6.36 The BlockRead interface

which provides split-phase commands to read from a block and compute CRCs. The log abstraction supports both circular and linear logs, but the choice must be made on a per-volume basis at compile-time:

```
generic configuration LogStorageC(volume_id_t volid, bool circular) {
  provides interface LogWrite;
  provides interface LogRead;
}
```

Listing 6.37 LogStorageC signature

6.5 Storage

The LogWrite interface provides split-phase erase, append and sync commands, and a command to obtain the current append offset:

```
interface LogWrite {
  command error_t append(void* buf, storage_len_t len);
  event void appendDone(void* buf, storage_len_t len, bool recordsLost,
         error_t error);

  command storage_cookie_t currentOffset();

  command error_t erase();
  event void eraseDone(error_t error);

  command error_t sync();
  event void syncDone(error_t error);
}
```

Listing 6.38 The LogWrite interface

However, many modules only need append: new logs start out empty, and logs are often intended to be persistent so are only rarely erased under programmer control, unlike an abstraction like block which is often erased before each use. Similarly, logs only really need to use sync either if the data is very important and must definitely not be lost, or if the mote is about to reboot, losing any cached data. For instance, FlashSampler's sample summarization code uses only append:

```
// in flashsampler.h:
enum { SUMMARY_SAMPLES = 256, // total samples in summary
       // downsampling factor: real samples per summary sample
       DFACTOR = TOTAL_SAMPLES / SUMMARY_SAMPLES };

module SummarizerC {
  provides interface Summary;
  uses interface BlockRead;
  uses interface LogWrite;
}
implementation
{
  uint16_t summary[SUMMARY_SAMPLES], samples[DFACTOR];
  uint16_t index; // of next summary sample to compute

  void nextSummarySample();

  command void Summary.summarize() {
    // Summarize the current sample block
    index = 0;

    nextSummarySample();
  }
```

```
void nextSummarySample() {
  // Read DFACTOR samples to compute the next summary sample
  call BlockRead.read(index * DFACTOR * sizeof(uint16_t), samples, sizeof samples);
}

event void BlockRead.readDone(storage_addr_t addr, void* buf, storage_len_t len,
                              error_t error) {
  // Average the DFACTOR samples which will become one summary sample
  uint32_t sum = 0;
  uint16_t i;

  for (i = 0; i < DFACTOR; i++) sum += samples[i];
  summary[index++] = sum / DFACTOR;

  // Move on to the next sample summary, or log the whole summary if we're done
  if (index < SUMMARY_SAMPLES)
    nextSummarySample();
  else
    call LogWrite.append(summary, sizeof summary);
}

event void LogWrite.appendDone(void* buf, storage_len_t len, bool recordsLost,
                               error_t error) {
  // Summary saved!
  signal Summary.summarized(error);
}

// Unused split-phase operations
event void BlockRead.computeCrcDone(storage_addr_t addr, storage_len_t len,
                                    uint16_t crc, error_t error) { }
event void LogWrite.eraseDone(error_t error) { }
event void LogWrite.syncDone(error_t error) { }
}
```

Listing 6.39 The LogWrite interface

Logs are read sequentially using the split-phase read command of the LogRead interface:

```
interface LogRead {
  command error_t read(void* buf, storage_len_t len);
  event void readDone(void* buf, storage_len_t len, error_t error);

  command storage_cookie_t currentOffset();

  command error_t seek(storage_cookie_t offset);
  event void seekDone(error_t error);
```

```
command storage_len_t getSize();
}
```

Listing 6.40 The LogRead interface

The log implementations guarantee that the data written by a single LogWrite.append operation will be either fully present, or fully absent, so we know that the log will not contain any partial summaries. So we could for instance read the log from the beginning, in 512-byte chunks (the size of one summary) and know that we have read a valid summary sample.

Log storage also provides limited seek support. The LogRead and LogWrite interfaces contain the following command

```
command storage_cookie_t currentOffset();
```

which returns a cookie that represents the position at which the next read or append will start. These cookies are *not* offsets in the log volume; they just contain enough information to allow the log storage implementation to find that position in the log. LogRead's split-phase seek command can seek to such cookies.

This would allow, e.g. the FlashSampler application, to report the position of the samples to a PC, and then reread specific samples on demand.

6.6 Exercises

1. Port the anti-theft application to a different platform than the micaz, by writing new theft detection code tailored to the particular sensors you have available.
2. Extend the anti-theft application so that motes notice the theft reports sent by neighboring motes and report them locally by blinking LED 1.
3. Extend the anti-theft application so that the user can select between the light-level (DarkC) and acceleration (MovingC) theft detection methods, and switch the LED 0 blinking on or off.
4. Add code to FlashSampler to transfer samples and sample summaries to the PC over the serial port. Allow the user to (re)request specific sample summaries or erase all the summaries.

7 Mote-PC communication

This chapter shows how to build a PC application that talks to motes. As we saw in Section 6.4, a PC typically interacts with a mote network by exchanging packets with a distinguished base station mote (occasionally, several motes) over a serial connection (Figure 6.2, page 95). The PC code in this chapter is written in Java, using the Java libraries and tools distributed with TinyOS. TinyOS also includes libraries and tool support for other languages (e.g. C). Please refer to the TinyOS documentation for more information on these other languages. The TinyOS Java code for communicating with motes is found under the net.tinyos package.

7.1 Basics

At the most basic level, PCs and motes exchange packets that are simply sequences of bytes, using a protocol inspired by, but not identical to, RFC 1663 [24] (more details on the protocol can be found in TEP 113 [7]). This packet exchange is not fully reliable: the integrity of packets is ensured by the use of a CRC, but invalid packets are simply dropped. Furthermore:

- Packets sent from a PC to a mote are acknowledged by the mote (but there is no retry if no acknowledge is received) – this prevents the PC from overloading the mote with packets.
- Packets sent from a mote to a PC are not acknowledged.

While it is possible to write mote communication code by reading and writing the raw bytes in packets, this approach is tedious and error-prone. Furthermore, any changes to the packet layout (e.g. adding a new field, changing the value of a constant used in the packets) requires corresponding changes throughout the code.

These problems are avoided on motes by using nesC's *external types* (Section 3.5.3) to specify packet layouts, and named constants to specify values. Two tools, *mig* and *ncg*, allow these packet layouts and constants to be used in PC programs: *mig* (for "message interface generator") generates code to encode and decode a byte array whose layout is specified by a nesC external type, and *ncg* (for "nesC constant generator") extracts named constants from a nesC program. Both tools generate code for multiple languages, including Java.

7.1.1 Serial communication stack

Like the radio stack (Section 6.3), the TinyOS serial stack follows a layered structure (Figure 6.1, page 89). The lowest level of the TinyOS serial communication stack is defined in TEP 113, and deals with the exchange of packets over an unreliable serial connection. Like radio packets, these are (in most cases) "active message" packets. In the rest of this chapter, we will use *message* to mean the payload of a serial AM packet, with different message types identifier by their AM type.

The layout of serial AM packets is defined using external types:

```
typedef nx_struct serial_header {
  nx_uint16_t dest;   /* packet destination */
  nx_uint16_t src;    /* packet sender */
  nx_uint8_t length;  /* payload length, in bytes */
  nx_uint8_t group;   /* identifies a specific sensor network */
  nx_uint8_t type;    /* AM type */
} serial_header_t;

typedef nx_struct serial_packet {
  serial_header_t header;
  nx_uint8_t data[];
} serial_packet_t;
```

Listing 7.1 Serial AM Packet layout

The type field is used to dispatch the payload (data[]). Serial AM packets have no footer. The dest (destination) and src (sender) fields are not directly relevant to packets exchanged over a serial connection, as the sender and destination are normally implicit. Similarly, group is not relevant: it is used to distinguish different motes participating in separate applications but sharing a physical connection. These fields are included for consistency with AM packets sent over a radio, and are used in some base station mote applications and ignored by others. For instance, in the BaseStation application that forwards radio AM packets to/from a PC (as serial AM packets):

- From BaseStation to the PC: src, dest and group are as in the received radio AM packet.
- From the PC to BaseStation: the transmitted radio AM packet has its destination set to dest, but sets its sender and group to those of the BaseStation mote.

With these choices, a PC with a BaseStation mote behaves much like a PC with a directly attached mote radio.

Conversely, the TestSerial application (part of the standard TinyOS distribution) ignores these three fields. TestSerial is an application designed to test serial connectivity: the mote and the PC simply send each other messages containing consecutive numbers. The mote displays the low-order three bits of these numbers on its LEDs, while the PC simply prints these numbers. TestSerial uses the following simple external type to

114 Mote-PC communication

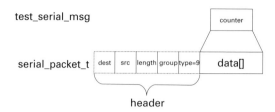

Figure 7.1 TestSerial packet layout.

specify its message layout:

```
typedef nx struct test_serial_msg {
  nx_uint16_t counter;
} test_serial_msg_t;
```

Listing 7.2 TestSerial packet layout

Serial AM packets containing a test_serial_msg_t message are identified by an AM type field of 9. Figure 7.1 shows the resulting connection between serial_packet_t and test_serial_msg_t external types (the figure does not show the lowest, packet-exchange layer as it is not specified using external types).

7.2 Using mig

The mig tool generates a class that represents a value of a given nesC **nx_struct** M.[1] This class provides methods to read, write, and report properties of the fields of M. Given a mig-generated class for M, the net.tinyos.message package provides classes and methods that allow you to send and receive serial AM packets whose payload is a value of type M.

Mig-generated classes represent structure values using a *backing array*, a byte array that stores the structure value in the same representation as it has on the mote, and hence the same representation as in TinyOS packets. Figure 7.2 shows a typical situation, with a mote sending a serial AM packet to a PC. On the mote, the packet is represented by x, a pointer to a serial_packet_t. The representation of x is the same as the representation of the packet on the serial connection to the PC. On the PC itself, a mig-generated class uses a backing array containing the same bytes to represent a serial_header_t (the external type for serial AM packet headers) value.

Mig-generated classes supports all features of C structures, including bitfields, arrays, unions, and nested structures. They allow you to perform all the operations (and more)

---

[1] It is also possible to use mig with regular C structures, but the resulting code is mote-specific, as different motes have different endianness and different rules for laying out C structures.

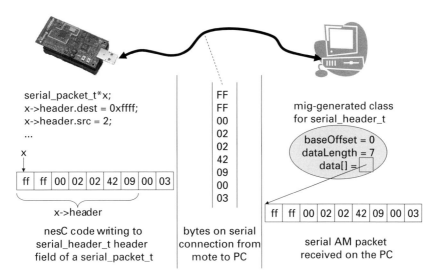

Figure 7.2 Mig and external types.

that you can do on structures in nesC programs:

- Create a new (all fields initialised to zero) structure value.
- Cast part of one structure to another (e.g. cast the data[] field of serial_packet_t to test_serial_msg_t).
- Read and write any field.
- Obtain information on a field: its offset, size, signedness, etc.
- Obtain the byte array representing a particular structure value.

The classes generated by mig are subclasses of net.tinyos.message.Message, to simplify writing generic code that can handle multiple message layouts. The main public function of the Message class is to provide access to the backing array of a mig-generated class. The backing array is a slice of a Java byte[] array, specified by an offset and length. This array, offset, and length are returned by the following three methods:

```
public byte[] dataGet();
public int baseOffset();
public int dataLength();
```

Listing 7.3 Backing array methods

For instance, in Figure 7.2, the mig-generated class for serial_header_t takes a 7-byte slice starting at offset 0 of the 9-byte array representing the whole serial AM packet. The use of a slice of a byte[] array allows a mig-generated class to represent a value stored within another mig-generated class, supporting the "casting" of part of one structure to another.

For test_serial_msg_t, the basic constructor and methods of the mig-generated TestSerialMsg class are:[2]

- `TestSerialMsg()`: create a new (fields initialised to zero) test_serial_msg_t value.
- `static int offset_counter()`: Return the offset (in bytes) of the counter field in a test_serial_msg_t structure.
- `get_counter()`: Return the value of the counter field.
- `set_counter(int value)`: Modify the value of the counter field.

7.2.1 Sending and receiving mig-generated packets

The net.tinyos.message package provides a class MoteIF (for "mote interface") that makes it simple to transmit and receive messages specified by a mig-generated class.

Before using MoteIF, you need to open a connection to your base station mote. This is done by calling the net.tinyos.packet.BuildSource.makePhoenix method, which takes as argument a *packet source* name. This packet source name specifies how you reach the base station mote (e.g. through a serial port) and all the relevant parameters (e.g. device name, baud rate, etc). In the rest of this section, we will use `"serial@COM1:telosb"` as our packet source, which denotes serial communication over serial port COM1: at the normal Telos mote baud rate (115 200 baud). Most Java programs that use MoteIF take a command-line argument that specifies the actual packet source. Section 7.4 discusses packet sources, and their motivation, in more detail.

Using MoteIF and mig, the code to transmit test_serial_msg_t messages is quite simple, as this simplified excerpt from the `TestSerial.java` file shows. Transmission of consecutive packets is performed using MoteIF's send method, which takes an object of a mig-generated class as argument:

```
public class TestSerial ... {
  private MoteIF moteIF;
  ...
  public void sendPackets() {
    int counter = 0;
    // Create uninitialized TestSerialMsg
    TestSerialMsg payload = new TestSerialMsg();

    try {
      while (true) {
        System.out.println("Sending packet " + counter);
        payload.set_counter(counter);
        moteIF.send(0, payload); // send payload to mote
        counter++;
        try {Thread.sleep(1000);}
        catch (InterruptedException exception) {}
      }
    }
```

---

[2] See the mig documentation for a complete description of the generated constructors and methods.

```
    catch (IOException exception) {
       System.err.println("Exception thrown when sending packets. Exiting.");
       System.err.println(exception);
    }
  }

  public static void main(String[] args) throws Exception {
    /* Open connection to the mote, and start sending packets */
    PhoenixSource phoenix = BuildSource.makePhoenix("serial@COM1:telosb", null);
    MoteIF mif = new MoteIF(phoenix);
    TestSerial serial = new TestSerial(mif);
    serial.sendPackets();
  }
}
```

Listing 7.4 Sending packets with mig and MoteIF

Received packets are handled following the AM "dispatch to a per-message handler" model. The registerListener method takes an object O of a mig-generated class and a handler as arguments. The object O specifies the AM type and layout, the handler must implement the net.tinyos.messages.MessageListener interface:

```
    public interface MessageListener {
       public void messageReceived(int to, Message m);
    }
```

Listing 7.5 Interface for handling received packets

When the handler is called, so is the destination address from the received packet, and m is a clone of O containing the payload of the received packet.

The resulting additions to TestSerial.java are simply to declare a messageReceived method, and register it as the handler for test_serial_msg_t in TestSerial's constructor:

```
public class TestSerial implements MessageListener {
  public TestSerial(MoteIF moteIF) {
    this.moteIF = moteIF;
    // Register this class as a handler for test_serial_msg_t AM packets
    this.moteIF.registerListener(new TestSerialMsg(), this);
  }

  public void messageReceived(int to, Message message) {
    // The actual type of 'message' is 'TestSerialMsg'
    TestSerialMsg msg = (TestSerialMsg)message;
    System.out.println("Received packet sequence number " + msg.get_counter());
  }
}
```

Listing 7.6 Receiving packets with mig and MoteIF

Note that the messageReceived method is called in a thread that is private to MoteIF. Your method must ensure that any accesses to shared data are appropriately synchronized with the rest of your program.

Both message transmission and reception need to know the AM type used to identify a particular message – transmission needs to set the AM type field (type in serial_packet_t), and reception needs to know which messages correspond to which AM type so that it can dispatch packets correctly. To support this, mig-generated classes contain an AM type field, accessed by the amType and amTypeSet methods. When mig generates a message class for structure X, it sets a default value for amType by looking for an **enum** constant named AM_X. For instance, the default AM type for **nx_struct** test_serial_msg is the value of AM_TEST_SERIAL_MSG (defined in `TestSerial.h`).

When there is no such constant in the source code, or when a message is used with several AM types, you must set the correct AM type using amTypeSet before calling the send or registerListener methods.

7.3 Using ncg

The ncg tool extracts **enum** constants from a nesC program or C header file, and generates a Java (or other language) file containing constants with the same name and value. This avoids specifying identical constants in two places (always a maintenance problem) when, e.g. your packets use "magic" values to represent commands, specify the size of arrays, etc.

Consider for instance the TinyOS Oscilloscope demo application where each mote periodically reports groups of sensor readings. Oscilloscope has a header file (`Oscilloscope.h`) that defines its message and associated constants:

```
enum {
  NREADINGS = 10,      /* Number of readings per message. */
  DEFAULT_INTERVAL = 256,  /* Default sampling period. */
  AM_OSCILLOSCOPE = 0x93
};

typedef nx_struct oscilloscope {
  nx_uint16_t version;   /* Version of the interval. */
  nx_uint16_t interval;  /* Sampling period. */
  nx_uint16_t id;        /* Mote id of sending mote. */
  nx_uint16_t count;     /* Readings start at count*NREADINGS */
  nx_uint16_t readings[NREADINGS];
} oscilloscope_t;
```

Listing 7.7 Constants and packet layout for Oscillscope application

Oscilloscope's Java GUI displays a graph of the sensor readings for each motes. To know the reading number (i.e. the position of readings on the graph X axis), it needs to know both the value of the count field and the value of the NREADINGS constants. It therefore uses a mig-generated class to decode received oscilloscope_t messages, and ncg to extract the NREADINGS value:

```
public class Constants {
    public static final byte NREADINGS = 10;
    ...
}
```

Listing 7.8 Class generated by ncg

The handler for oscilloscope_t messages can now fetch the readings array from the received message, and store its values in the Java data array at offset `count * NREADINGS`:

```
public void messageReceived(int to, Message msg) {
  OscilloscopeMsg omsg = (OscilloscopeMsg)msg;
  int start = omsg.get_count() * Constants.NREADINGS;
  // get_readings returns a Java array with elements corresponding to
  // the values in the nesC array
  int readings[] = omsg.get_readings();

  for (int i = 0; i < readings.length; i++)
      data[start + i] = readings[i];
}
```

Listing 7.9 Simplified code to save received samples

The implementation of ncg relies on running the nesC compiler in a special mode, and is thus able to extract the values of enum constants defined by expressions and uses of nesC's special unique and uniqueCount functions (see Chapter 9.1), not just simple constants. However, as a result, ncg also has two limitations:

- It cannot extract the values of **#define** constants, as these are handled by the C preprocessor and invisible to ncg.
- If the header file or nesC program passed to ncg has compile errors, then ncg will not produce any output.

7.4 Packet sources

The Java mote interface uses packet sources to abstract the different ways to talk to a mote. These include:

- direct serial connections
- remote (over Ethernet) serial connections
- connections to simulated motes
- connection via serial forwarders, which allow several programs to talk to the same mote

A packet source has the form *connection@arguments*, where *connection* is a word that specifies the kind of connection (serial for direct serial connections, sf for serial forwarders, etc.) and *arguments* is the arguments needed by the connection (e.g. serial port device name and baud rate as in `serial@COM1:115200`). Executing

```
java net.tinyos.packet.BuildSource
```

prints a summary of all packet sources and their arguments. As mentioned above, most Java programs have some way of specifying the packet source that should be used to connect to the base station mote, e.g. an environment variable (usually MOTECOM), a command-line argument, or a field in a GUI.

One of the more important packet sources is the *serial forwarder*. The serial forwarder is a Java program[3] that connects to a mote and forwards packets to and from its (multiple) clients. These clients connect to the serial forwarder via TCP/IP. Using a serial forwarder serves multiple purposes:

- Allow remote access to a base station mote, so that, e.g. a GUI can access a sensor network from a remote location.
- Allow both a regular application and a debugging system to talk to the same base station mote. For instance, it's often useful to simply display all packets received from a mote (using the net.tinyos.tools.Listen Java program) without having to change your code.
- Split your application into multiple independent parts, each of which can talk to the mote. For instance, the `TestSerial.java` application could be split into separate receive and transmit applications which would talk independently to the mote via a serial forwarder.

7.5 Example: simple reliable transmission

As we saw at the beginning of this chapter, the basic mote-PC communication infrastructure is unreliable: packets are checked for integrity but there is no retransmission in case of failure. In this section, we show how to build a simple reliable and reusable message transmission layer, and use it to build a reliable version of the TestSerial application. The focus of this section is on showing how to build a generic, reusable mote communication layer in Java.

The code for this simple reliable transmission protocol can be found in the "TinyOS Programming" section of TinyOS's contributed code directory (Section 1.5), under the name ReliableSerial. This directory contains the Java (ReliableMoteIF class) and nesC (ReliableSerialC component) implementations of this protocol, and ReliableTestSerial, a version of the nesC TestSerial application built over ReliableSerialC.

---

[3] There are also C and C++ implementations.

7.5 Example: simple reliable transmission

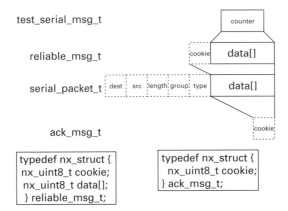

Figure 7.3 Reliable transmission packet layouts.

7.5.1 Reliable transmission protocol

The reliable message transmission protocol is very simple. Messages are sent with a "cookie" (changed on every transmission) and are repeatedly sent (after a timeout) until an acknowledgement with the same cookie is received. Transmissions are not overlapped: transmission of *message2* does not start until *message1* has been acknowledged.

On the reception side, duplicates (detected by the presence of an identical cookie) are suppressed before passing the message on to the upper layer. However, even duplicate messages are acknowledged, as it is possible that the earlier acknowledgements were lost.

In concrete terms, two kinds of packets are used by this protocol: a message transmission packet (reliable_msg_t) and an acknowledgement packet (ack_msg_t). These packets are shown in Figure 7.3, along with the TestSerial messages that are transmitted by the ReliableTestSerial application.

7.5.2 Reliable transmission in Java

The ReliableMoteIF class contains the Java implementation of the simple reliable message transmission protocol. It uses mig to generate AckMsg and ReliableMsg so that it can build and decode ack_msg_t and reliable_msg_t packets respectively. ReliableMoteIF uses ncg to access the retransmission timeout (ACK_MSG_TIMEOUT) specified in the nesC implementation (the ReliableSerialC component).

This class is implementing a generic reusable layer, so should be capable of transmitting and receiving an arbitrary message whose layout is specified by a mig-generated class. As a result, the interface is so similar to that of MoteIF that we simply make it a subclass of MoteIF. We first show the transmission side:

```
import net.tinyos.packet.*;
import net.tinyos.message.*;
import java.io.*;
```

```java
class ReliableMoteIF extends MoteIF {
    /* Build an object for performing reliable transmission via 'base' */
    public ReliableMoteIF(PhoenixSource base) {
      super(base);
      /* Register handler ('ackMsgReceived' method below) for receiving acks */
      super.registerListener
         (new AckMsg(), new MessageListener() {
             public void messageReceived(int to, Message m) {
             ackMsgReceived((AckMsg)m);
          }
         });
  }

  /* Send side */
  /* --------- */
  private Object ackLock = new Object(); /* For synchronization with ack handler */
  private short sendCookie = 1; /* Next cookie to use in transmission */
  private boolean acked; /* Set by ack handler when ack received */
  private final static int offsetUserData = ReliableMsg.offset_data(0);

  /* Send message 'm' reliably with destination 'to' */
  public void send(int to, Message m) throws IOException {
     synchronized (ackLock) {
     /* Build a reliable_msg_t packet with the current cookie and the payload in m,
      * total packet size is 'm.dataLength() + offsetUserData' */
     ReliableMsg rmsg = new ReliableMsg(m.dataLength() + offsetUserData);
     rmsg.set_cookie(sendCookie);
     System.arraycopy(m.dataGet(), m.baseOffset(),
                      rmsg.dataGet(), offsetUserData,
                      m.dataLength());

       /*Repeatedly transmit 'rmsg' until the ack handler tells us an ack is received.*/
         acked = false;
         for (;;) {
            super.send(to, rmsg);
            try {
                ackLock.wait(RelConstants.ACK_TIMEOUT);
            }
            catch (InterruptedException e) { }
            if (acked)
                break;
            System.err.printf("retry\n");
      }
      /* Pick a new cookie for the next transmission */
      sendCookie = (short)((sendCookie * 3) & 0xff);
      }
  }
```

7.5 Example: simple reliable transmission

```java
    /* Handler for ack messages. If we see an ack for the current transmission,
     * notify the 'send' method. */
    void ackMsgReceived(AckMsg m) {
        synchronized (ackLock) {
            if (m.get_cookie() == sendCookie) {
                acked = true;
                ackLock.notify();
            }
        }
    }

    ... receive side ...
}
```

Listing 7.10 Reliable transmission protocol in Java - transmission

The send method is synchronized to enforce the rule that a message must be acknowledged before the next message is sent.

This code relies on the fact that the ackMsgReceived handler executes in a different thread than send. After sending a message, the send thread goes to sleep for ACK_MSG_TIMEOUT ms. When an acknowledgement is received, ackMsgReceived checks the cookie against that of the message being transmitted. If it is equal, acked is set to true and the send thread is woken up. If no acknowledgement is received within ACK_MSG_TIMEOUT ms, send will resend the message and wait again. Note that acknowledgements received when send is not waiting are effectively ignored, as they should be.

After transmission is successful, send picks a new cookie – we multiply the old value by three just to make it clear that these cookies are not sequence numbers.

The embedding of the user's message in a reliable_msg_t packet is simply done by using System.arrayCopy to copy data between the backing arrays of the user's and reliable_msg_t's mig-generated classes. Note also how the mig-generated offset_data method is used to find the payload offset in reliable_msg_t (see the offsetUserData constant). This code is very similar to MoteIF's embedding of user messages in serial packets.

Reliable reception is implemented by overriding MoteIF.registerListener. Note, however, that our reliable transmission protocol only knows about one message type (it has no equivalent to AM's type field), so you can only actually register a listener for one message type:

```java
class ReliableMoteIF extends MoteIF {
    ...
    /* Receive side */
    /* ------------ */
    private Message template;
    private MessageListener listener;
```

```java
    /* Build a reliable receive handler for 'template' messages */
    public void registerListener(Message m, MessageListener l) {
        template = m;
        listener = l;

        /* Register handler (reliableMsgReceived method below) for receiving
         * reliable_msg_t messages */
        super.registerListener
            (new ReliableMsg(),
             new MessageListener() {
                public void messageReceived(int to, Message m) {
                    reliableMsgReceived(to, (ReliableMsg)m);
                }
            });
    }

    private short recvCookie; /* Cookie of last received message */

    void reliableMsgReceived(int to, ReliableMsg rmsg) {
        /* Acknowledge all received messages */
        AckMsg ack = new AckMsg();
        ack.set_cookie(rmsg.get_cookie());
        try {
            super.send(MoteIF.TOS_BCAST_ADDR, ack);
        }
        catch (IOException e) {
            /* The sender will retry and we'll re-ack if the send failed */
        }

        /* Don't notify user of duplicate messages */
        if (rmsg.get_cookie() != recvCookie) {
            recvCookie = rmsg.get_cookie();

            /* Extract payload from 'rmsg' and copy it into a copy of 'template'.
             * The payload is all the data in 'rmsg' from 'offsetUserData' on */
            Message userMsg = template.clone(rmsg.dataLength() - offsetUserData);
            System.arraycopy(rmsg.dataGet(), rmsg.baseOffset() + offsetUserData,
                             userMsg.dataGet(), 0,
                             rmsg.dataLength() - offsetUserData);
            listener.messageReceived(to, userMsg);
        }
    }
}
```

Listing 7.11 Reliable transmission protocol in Java - transmission

The protocol implementation part of reliableMsgReceived is straightforward: it simply acknowledges all received messages, and ignores consecutive messages with the same cookie. Its main complexity is in extracting the payload from received reliable_msg_t packets and building a message that follows the user's template object. This task is accomplished by using the `clone(n)` method of Message, which makes a copy of a message with a new backing array. The data from the payload portion of the backing array of the received reliable_msg_t packet is then simply copied over to this new backing array. Again, this code is very similar to that found in MoteIF to extract the payload from a received serial packet.

7.5.3 Reimplementing TestSerial

Switching TestSerial's Java code to use the new reliable transmission layer is very straightforward because ReliableMoteIF is a subclass of MoteIF. We just construct a ReliableMoteIF object in main:

```
public static void main(String[] args) throws Exception {
  /* Open connection to the mote, and start sending packets */
  PhoenixSource phoenix=BuildSource.makePhoenix("serial@COM1:telosb", null);
  MoteIF mif = new ReliableMoteIF(phoenix);
  TestSerial serial = new TestSerial(mif);
  serial.sendPackets();
}
```

Listing 7.12 A reliable TestSerial.java

7.6 Exercises

1. ReliableMoteIF is not a proper replacement for MoteIF because our reliable transmission protocol only supports a single message type. Extend reliable_msg_t, ReliableMoteIF, and ReliableSerialC (the nesC reliable protocol implementation) with an AM-type-like dispatch field.
2. Implement a simple PC application that talks to the AntiTheft base station mote of Section 6.4.4, to print theft report and allow the user to change theft-detection settings.
3. Improve the performance of the reliable transmission protocol by allowing multiple outstanding un-acknowledged packets using, e.g. a windowed protocol.

Part III

Advanced programming

8 Advanced components

Chapter 3 introduced components to a level of detail sufficient for building applications. This chapter presents more advanced component topics than implementing services typically requires, such as writing generic components and using parameterized interfaces.

8.1 Generic components review

Generic components (introduced in Section 3.3.4) provide code reuse through a code-copying mechanism: each instance of a generic component is effectively a new (non-generic) component with the parameter values substituted in. For instance, instantiating BitVectorC

```
generic module BitVectorC(uint16_t maxBits) {
  provides interface Init;
  provides interface BitVector;
}
implementation {
  uint8_t bits[(maxBits + 7) / 8];
  ...
}
```

with

```
configuration MyAppC { }
implementation {
  components new BitVectorC(77) as MyVector, MyAppP;

  MyAppP.BitVector -> MyVector;
}
```

is effectively the same as creating a copy of BitVectorC.nc with maxBits=77:

```
module MyVector {
  provides interface Init;
  provides interface BitVector;
}
```

```
implementation {
  uint8_t bits[(77 + 7) / 8];
  ...
}
```

Generic components use a code-copying approach for two reasons: simplicity and types. If generic modules did not use a code-copying approach, then there would be a single copy of the code that works for all instances of the component. This is difficult when a generic component can take a type as an argument, as allocation size, offsets, and other considerations can make a truly single copy infeasible. Even non-type arguments can create such problems, as they can be used to specify array sizes, switch case values, etc. For similar reasons, C++ templates create a copy of code for each different set of template arguments.

Finally, even when it would be possible to share code across instances, it would require adding an argument, similar to the `this` pointer of object-oriented languages, to all of the functions. This argument would indicate which instance is executing. Additionally, all variable accesses would have to offset from this pointer. In essence, the execution time and costs of functions might change significantly (offsets rather than constant accesses). In order to provide simple, easy to understand and run-time efficient components, nesC uses a code-copying approach, sacrificing possible reductions in code size.

Code-copying applies to configurations as well as modules. Copying a module copies executable code and variables into the final application. Copying a configuration copies component and wiring statements, possibly leading to further component instantiations:

```
generic configuration DoubleBitVectorC(uint16_t maxBits) {
  provides interface Init;
  provides interface BitVector as Bits1;
  provides interface BitVector as Bits2;
}
implementation {
  components new BitVector(maxBits) as BV1;
  components new BitVector(maxBits) as BV2;

  Init = BV1; // Calling Init.init initialises
  Init = BV2; // both bitvectors
  Bits1 = BV1;
  Bits2 = BV2;
}
```

Listing 8.1 Instantiation within a generic configuration

Instantiating DoubleBitVectorC will create, and wire, two bit vectors. In summary, a generic module defines a piece of repeatable executable logic, while a generic configuration defines a repeatable pattern of composition between components. As Chapter 9 will show, generic configurations can be very powerful and are often the most intricate part of service implementations.

Unlike standard components, generics can only be named by the configuration that instantiates them. For example, in the case of MyAppC, no other component can wire to the BitVectorC that it instantiates. The generic is private to MyAppC. The only way it can be made accessible is to wire its interfaces – in the case of MyAppC, the bit vector is made accessible to MyAppP only. One way to make an instance generally available is to create a singleton configuration and exporting the necessary interfaces. For example, let's say you needed a bit vector to keep track of which of 37 system services are running or not. You want many components to be able to access this vector, but BitVectorC is a generic. So you write a component like this:

```
configuration SystemServiceVectorC {
  provides interface BitVector;
}
implementation {
  components MainC, new BitVectorC(37);
  MainC.SoftwareInit -> BitVectorC;
  BitVector = BitVectorC;
}
```

Listing 8.2 The fictional component SystemServiceVectorC

Now many components can refer to this particular bit vector. SystemServiceVectorC could have exported BitVectorC.Init rather than wiring it to MainC, but that would have just required this wiring to be performed elsewhere. While you can make a singleton out of a generic by instantiating it within one, the opposite is not true: a component is either instantiable or not.

8.2 Writing generic modules

As the examples above showed, the body of a generic module (configuration) is simply a module (configuration) which can use the generic's arguments as types, constants, etc. Another example is the queue implementation, QueueC:

```
generic module QueueC(typedef queue_t, uint8_t queueSize) {
  provides interface Queue<queue_t>;
}
implementation {
  queue_t queue[queueSize];
  uint8_t head = 0, tail = 0, size = 0;

  command queue_t Queue.head() {
    return queue[head];
  }
  ...
```

Listing 8.3 QueueC excerpt

QueueC uses queue_t like a regular C type, declaring an array (queue) and returning a queue_t value (Queue.head). The queueSize parameter behaves like a C constant, so can be used to size the queue array.

8.2.1 Type arguments

Unlike C++ templates, generic components are type-checked when they are declared. By default, the only operations allowed on type arguments are declaring variables and copying values. For instance, you can do

```
queue_t x, y;
x = y;
```

but not

```
queue_t x, y;
x = y + 1;
```

as nesC would not know what this meant if, e.g. queue_t was a C structure.

However, some components have type arguments that only make sense as some kind of integer (signed 8-bit integer, unsigned 32-bit integer, etc). For those cases, generic component type parameters can be suffixed with @integer() to allow the use of integer operations on values of that type. For instance, this ConstantSensorC component is a generic implementation of Read returning a constant value:

```
generic module ConstantSensorC(typedef width_t @integer(), uint32_t val) {
  provides interface Read<width_t>;
}
implementation
{
  task void senseResult() {
    signal Read.readDone(SUCCESS, (width_t)val);
  }

  command error_t Read.read() {
    return post senseResult();
  }
}
```

Listing 8.4 A generic constant sensor

The type of values returned by ConstantSensorC's Read interface is width_t, and the constant returned is the integer val. The @integer() is necessary to make the cast to width_t in senseResult legal: an integral value like val *can* be cast to any integer type, but not to some unknown type.[1]

---

[1] Note that removing the cast from the source code would not change anything, as the cast is implied by passing val to Read.readDone, whose second argument is of type width_t.

Instantiations of ConstantSensorC must use an integer type for width_t:

```
components new ConstantSensorC(uint16_t, 23) as C1; // legal
components new ConstantSensorC(struct X, 39) as C2; // compile-time error
```

There is also an @number() suffix, that restricts the type parameter to an integer or floating-point type, and the operations to those legal on all such types. The @integer() and @number() suffixes are other examples of nesC's attributes, which we saw earlier in the declaration of combine functions (Section 4.4.3) and which will be covered in more detail in Section 8.4.

8.2.2 Abstract data types as generics

Abstract data types (ADTs) in TinyOS are usually represented in one of two ways: as generic modules or through an interface with commands that take by-reference arguments. With a module, the ADT state is stored in module variables and accessed by commands from provided interfaces. For example, many TinyOS components access queues through the Queue typed interface:

```
interface Queue<t> {
  command bool empty();
  command uint8_t size();
  command uint8_t maxSize();
  command t head();
  command t dequeue();
  command error_t enqueue(t newVal);
  command t element(uint8_t idx);
}
```

Listing 8.5 Queue interface (repeated)

The Queue ADT implementation, QueueC, takes the queue type and size as parameters:

```
generic module QueueC(typedef queue_t, uint8_t queueSize) {
  provides interface Queue<queue_t>;
}
implementation {
  queue_t queue[queueSize];
  uint8_t head = 0, tail = 0, size = 0;

  command bool Queue.empty() {
    return size == 0;
  }

  command uint8_t Queue.size() {
    return size;
  }
```

```
command uint8_t Queue.maxSize() {
  return queueSize;
}

command queue_t Queue.head() {
  return queue[head];
}

command queue_t Queue.dequeue() {
  queue_t t = call Queue.head();
  if (!call Queue.empty()) {
    if (++head == queueSize) head = 0;
    size--;
  }
  return t;
}

command error_t Queue.enqueue(queue_t newVal) {
  if (call Queue.size() < call Queue.maxSize()) {
    queue[tail++] = newVal;
    if (tail == queueSize) tail = 0;
    size++;
    return SUCCESS;
  }
  else {
    return FAIL;
  }
}

command queue_t Queue.element(uint8_t idx) {
  idx += head;
  if (idx >= queueSize) idx -= queueSize;
  return queue[idx];
}
}
```

Listing 8.6 QueueC implementation

Consistency issues can arise when multiple components share an ADT. In general, an abstract data type implementation's commands should behave correctly even if there are multiple clients. However, it is up to the clients to handle consistency issues across multiple calls: if client A calls Queue.size in two different tasks and client B calls Queue.enqueue, then client A may get different sizes, depending on the order of the calls. These issues are more complex for abstract data types such as BitVectorC that can be used from asynchronous code – see Chapter 11 for more discussion on this topic.

8.2.3 ADTs in TinyOS 1.x

TinyOS 1.x uses only the second approach, an interface with by-reference parameters, because it does not have generic modules. For example, the Maté virtual machine supports scripting languages with typed variables, and provides functionality for checking and setting types. In this case, the ADT is a script variable. In the interface MateTypes below, a MateContext* is a thread and a MateStackVariable* is a variable:

```
interface MateTypes {
  command bool checkTypes(MateContext* context, MateStackVariable* var, uint8_t type);
  command bool checkMatch(MateContext* context, MateStackVariable* v1, MateStackVariable* v2);
  command bool checkValue(MateContext* context, MateStackVariable* var);
  command bool checkInteger(MateContext* context, MateStackVariable* var);
  command bool isInteger(MateContext* context, MateStackVariable* var);
  command bool isValue(MateContext* context, MateStackVariable* var);
  command bool isType(MateContext* context, MateStackVariable* var, uint8_t type);
}
```

Listing 8.7 Representing an ADT through an interface in TinyOS 1.x

The MateTypes interface is provided by a non-generic module (MTypeCheck) that accesses, and the actual MateStackVariable objects, are declared in various other modules. The consistency issues are similar to when ADTs are provided by generic modules: if two clients use an ADT interface to access the same variable, then the ADT should ensure each operation behaves consistently, and clients should deal with any higher-level consistency issues.

8.3 Parameterized interfaces

One way to provide multiple independent instances of the same interface is to implement the interface in a generic module and instantiate it multiple times. However, this has two drawbacks. First, it wastes code space. Second, in many cases, the multiple instances are not really fully independent, but need to perform some degree of cooperation. For instance, the TinyOS timer component (HilTimerMilliC) *virtualizes* a platform's hardware timer: it implements multiple logical timers on top of a single hardware timer. From the application programmer's perspective there are, e.g. 10 independent Timer interfaces, but the timer implementation needs to set the single hardware timer to the earliest deadline amongst these 10 timers.

One way HilTimerMilliC could provide multiple timers is by having a long signature:

```
configuration HilTimerMilliC {
  provides interface Timer<TMilli> as Timer0;
  provides interface Timer<TMilli> as Timer1;
  provides interface Timer<TMilli> as Timer2;
  provides interface Timer<TMilli> as Timer3;
  ...
  provides interface Timer<TMilli> as Timer100;
}
```

Listing 8.8 Timers without parameterized interfaces

While this works, it is somewhat painful and leads to a lot of repeated code. Every instance needs to have its own implementation. That is, there will be 100 different startPeriodic functions, even though they're almost completely identical. Another approach could be to have a call parameter to the Timer interface that specifies which timer is being changed, sort of like a file descriptor in POSIX file system calls. In this case, HilTimerMilliC would look like this

```
configuration HilTimerMilliC {
  provides interface Timer;
}
```

Listing 8.9 Timers with a single interface

Components that use timers would have some way of generating unique timer identifiers, and would pass them in every call:

```
call Timer.startPeriodic(timerDescriptor, 1024); // Fire at 1Hz
```

While this approach works – it doesn't lead to multiple implementations – passing the parameter is generally unnecessary, in that components generally allocate some number of timers and then only use those timers. That is, the set of timers a component uses – and the size of the set – are generally known at compile-time. Making the caller pass the parameter at run-time is therefore unnecessary, and could possibly introduce bugs (e.g. if the wrong descriptor is passed).

There are other situations when a component wants to provide a large number of interfaces, such as communication. Active messages have an 8-bit type field, which is essentially a protocol identifier (Section 6.3). In the Internet, the valid protocol identifiers for IP are well specified,[2] and many port numbers for TCP are well established. When a node receives an IP packet with protocol identifier 6, it knows that this is a TCP packet and dispatches it to the TCP stack. Active messages need to perform a similar function, albeit without the standardization of IANA: a network protocol needs to be able to register to send and receive certain AM types. Like timers, with basic interfaces there are two ways to approach this: code redundancy or run-time parameters. That is, to handle AM type 15, you could either have a packet layer providing 256 named Send interfaces:

```
components NetworkProtocolP, PacketLayerC;
NetworkProtocolP.Send -> PacketLayerC.Send15;
```

or the network protocol code could look like this:

```
call Send.send(15, msg, sizeof(payload_t));
```

Neither of these solutions is very appealing. The first leads to a lot of redundant code, wasting code memory. Furthermore, the use of hardwired names can easily lead to maintenance problems and/or bugs: returning to the timers, if a sensor filter and a routing stack both wire to Timer3, there's no way to separate them without changing

---

[2] www.iana.org/assignments/protocol-numbers

the code text of one of them to read "Timer4." One way to manage the namespace would be to have components leave their timers unwired and then expect the application to resolve all of them. But this places a large burden on an application developer. For example, a small application that builds on top of a lot of large libraries might have to wire eight different timers. Additionally, this approach leads to system components that are not self-contained, working abstractions: they have remaining dependencies that an application developer needs to resolve.

The second approach (passing extra arguments) is superior to the first at first glance, but it turns out to have even more significant problems, especially in the context of nesC's component model. First, in many cases the identifier is known at compile-time. Requiring the caller to pass it as a run-time parameter is unnecessary and is a possible source of bugs. Second, and more importantly, it pushes identifier management into the caller. For example, returning again to timers:

```
call Timer.startPeriodic(timerDescriptor, 1024); // Fire at 1Hz
```

From the calling component's perspective, it doesn't care which timer it's using. All it cares is that it has its own timer. Making the identifier part of the call forces the module to know (and manage) the identifier name. The third and largest problem, however, is not with calls out to other components but with calls in from other components. In Timer, for example, how does the timer service signal a fired() event to the correct component? Because the identifier is a run-time parameter, the only way is for Timer.fired() to fan-out to all timers, and have them all check the identifier.

8.3.1 Parameterized interfaces and configurations

To support abstractions that provide sets of interfaces, nesC has parameterized interfaces. A parameterized interface is essentially an array of interfaces, and the array index is the parameter. For example, this is the signature of HilTimerMilliC:

```
configuration HilTimerMilliC {
  provides interface Init;
  provides interface Timer<TMilli> as TimerMilli[uint8_t num];
  provides interface LocalTime<TMilli>;
}
```

Listing 8.10 HilTimerMilliC signature

HilTimerMilliC is a platform-specific configuration that presents a millisecond granularity timer stack. By providing 256 separate instances of Timer, HilTimerMilliC can support up to 256 independent timers. Calling Timer.startPeriodic on one interface instance will cause that instance to signal Timer.fired events. Normally, components don't wire directly to HilTimerMilliC. Instead, they use TimerMilliC, which presents a simpler interface (see Section 6.1).

Packet communication is another example use of parameterized interfaces. ActiveMessageC is a platform-specific configuration for single-hop packet communication:

```
configuration ActiveMessageC {
  provides {
    interface Init;
    interface SplitControl;

    interface AMSend[uint8_t id];
    interface Receive[uint8_t id];
    interface Receive as Snoop[uint8_t id];

    interface Packet;
    interface AMPacket;
    interface PacketAcknowledgements;
  }
}
```

Listing 8.11 ActiveMessageC signature

AMSend, Receive, and Snoop are all parameterized interfaces. Their parameter is the AM type of the message (the protocol identifier). Normally, components don't wire directly to ActiveMessageC. Instead, they use AMSenderC, AMReceiverC, which present a simpler interface (see Section 6.3). However, some test applications for the basic AM abstraction, such as TestAM, use ActiveMessageC directly. The module TestAMC sends and receives packets:

```
module TestAMC {
  uses {
    ...
    interface Receive;
    interface AMSend;
    ...
  }
}
```

Listing 8.12 Signature of TestAMC

TestAMAppC is the configuration that wires up the TestAMC module:

```
configuration TestAMAppC {}
implementation {
  components MainC, TestAMC as App;
  components ActiveMessageC;
```

```
    MainC.SoftwareInit -> ActiveMessageC;
    App.Receive -> ActiveMessageC.Receive[240];
    App.AMSend -> ActiveMessageC.AMSend[240];
    ...
}
```

Listing 8.13 Wiring TestAMC to ActiveMessageC

Note that TestAM has to wire SoftwareInit to ActiveMessageC because it doesn't use the standard abstractions, which auto-wire it. This configuration means that when TestAMC calls AMSend.send, it calls ActiveMessageC.AMSend number 240, so sends packets with protocol identifier 240. Similarly, TestAMC receives packets with protocol identifier 240. Because these constants are specified in the configuration, they are not bound in the module: from the module's perspective, they don't even exist. That is, from TestAMC's perspective, these two lines of code are identical:

```
TestAMC.AMSend -> ActiveMessageC.AMSend240;  // Not real TinyOS code
TestAMC.AMSend -> ActiveMessageC.AMSend[240];
```

The difference lies in the component with the parameterized interface. The parameter is essentially another argument in functions of that interface. In ActiveMessageC.AMSend, for example, the parameter is an argument passed to it in calls to send() and which it must pass in signals of sendDone(). But the parameterized interface gives you two key things. First, it automatically fills in this parameter when TestAMC calls send (nesC generates a stub function to do so, and inlining makes the cost negligible). Second, it automatically dispatches on the parameter when ActiveMessageC signals sendDone (nesC generates a switch table based on the identifier).

8.3.2 Parameterized interfaces and modules

In reality, ActiveMessageC is a configuration that encapsulates a particular chip's radio stack, and that stack may itself be a configuration. For instance, for the CC2420 radio, ActiveMessageC encapsulates the CC2420ActiveMessageC configuration, which itself encapsulates the CC2420ActiveMessageP module:

```
module CC2420ActiveMessageP {
  provides {
    interface AMSend[am_id_t id];
    ...
  }
}
```

Listing 8.14 A possible module underneath ActiveMessageC

Within CC2420ActiveMessageP, this is what the parameterized interface looks like:

```
command error_t AMSend.send[am_id_t id](am_addr_t addr, message_t* msg, uint8_t len) {
  cc2420_header_t* header = getHeader( msg );
  header->type = id;
  ...
}
```

Listing 8.15 Parameterized interface syntax

The interface parameter precedes the function argument list, and the implementation can treat it like any other argument. Basically, it is a function argument that the nesC compiler fills in when components are composed. When CC2420ActiveMessageP wants to signal sendDone, it pulls the protocol identifier back out of the packet and uses that as the interface parameter:

```
event void SubSend.sendDone(message_t* msg, error_t result) {
  signal AMSend.sendDone[call AMPacket.type(msg)](msg, result);
}
```

Listing 8.16 Dispatching on a parameterized interface

If the AM type of the packet is 240, then the dispatch code nesC generates will cause this line of code to signal the sendDone wired to ActiveMessageC's AMSend[240] interface, which in this case is ultimately wired to TestAMC's AMSend.sendDone.

CC2420ActiveMessageP.Receive looks similar to sendDone. The AM implementation receives a packet from a lower-level component and dispatches on the AM type to deliver it to the correct component. Depending on whether the packet is destined to the local node, it signals either Receive.receive or Snoop.receive:

```
event message_t* SubReceive.receive(message_t* msg, void* payload, uint8_t len) {
  if (call AMPacket.isForMe(msg)) {
    return signal Receive.receive[call AMPacket.type(msg)](msg, payload, len - CC2420_SIZE);
  }
  else {
    return signal Snoop.receive[call AMPacket.type(msg)](msg, payload, len - CC2420_SIZE);
  }
}
```

Listing 8.17 How active message implementations decide on whether to signal to Receive or Snoop

The subtraction of CC2420_SIZE is because the lower layer has reported the entire size of the packet, while to layers above AM the size of the packet is only the data payload (the entire size minus the size of headers and footers, that is, CC2420_SIZE).

Parameterized interfaces get the best of both worlds. Unlike the name-based approach (e.g. Send240) described above, there is a single implementation of the call. Additionally, since the parameter is a value, unlike a name it can be configured and set. For example, a component can do something like this:

```
#ifndef ROUTING_TYPE
#define ROUTING_TYPE 201
#endif

RouterP.AMSend -> PacketSenderC.AMSend[ROUTING_TYPE];
```

Listing 8.18 Defining a parameter

Pushing constants into wiring avoids the pitfalls of using run-time parameters. Because the constant is set at compile-time, nesC can automatically fill it in and dispatch based on it simplifying the code and improving the efficiency of outgoing function invocations.

Note that you can also wire entire parameterized interfaces:

```
configuration CC2420ActiveMessageC {
  provides interface AMSend[am_id_t id];
} {...}
configuration ActiveMessageC {
  provides interface AMSend[am_id_t id];
}
implementation {
  components CC2420ActiveMessageC;
  AMSend = CC2420ActiveMessageC;
}
```

Listing 8.19 Wiring full parameterized interface sets

Programming Hint 18 USE A PARAMETERIZED INTERFACE WHEN YOU NEED TO DISTINGUISH CALLERS OR WHEN YOU HAVE A COMPILE-TIME CONSTANT PARAMETER.

8.3.3 Defaults

Because a module's call points are resolved in configurations, a common compile error in nesC is to forget to wire something. The equivalent in C is to forget to include a library in the link path, and in Java it's to include a jar. Usually, a dangling wire represents a bug in the program. With parameterized interfaces, however, often they don't.

Take, for example, the Receive interface of ActiveMessageC. Most applications receive a few AM types, maybe 15 at most: they don't respond to or use every protocol ever developed. However, there's this call in CC2420ActiveMessageP:

```
return signal Receive.receive[call AMPacket.type(msg)]
(msg, payload, len - CC2420_SIZE);
```

On one hand, if all of the nodes in the network run the same executable, it's possible that none of them will ever send a packet of, say, AM type 144. However, if there are other nodes nearby, or if packets are corrupted but still pass a CRC check (this is rare, but does happen), then it's very possible that a node which doesn't care about protocol 144 will receive a packet of this type. Therefore nesC expects the receive event to have a handler: it needs to execute a function when this happens. But the application doesn't wire to Receive[144], and making a developer wire to all of the unwired instances is unreasonable, especially as they're all null functions (in the case of Receive.receive, the handler just returns the packet passed to it).

To avoid this problem, nesC provides default handlers. A default handler is an implementation of a function that's used if no implementation is wired in. If a component wires to the interface, then that implementation is used. Otherwise, the call (or signal) goes to the default handler. For example, CC2420ActiveMessageP has the following default handlers:

```
default event message_t* Receive.receive[am_id_t id](message_t* msg, void* p, uint8_t len) {
  return msg;
}

default event message_t* Snoop.receive[am_id_t id](message_t* msg, void* p, uint8_t len) {
  return msg;
}

default event void AMSend.sendDone[uint8_t id](message_t* msg, error_t err) {
  return;
}
```

Listing 8.20 Default events in an active message implementation

In the TestAM application, TestAMAppC wires TestAMC to ActiveMessageC.Receive[240]. Therefore, on the Telos or micaz platform, when the radio receives a packet of AM type 240, it signals TestAMC's Receive.receive. Since the application doesn't use any other protocols, when it receives an active message of any other type it signals CC2420ActiveMessageP's default handler.

Default handlers can be dangerous, as they circumvent compile-time checks that component interfaces are connected. Using them carelessly can cause code to work improperly. For example, while CC2420ActiveMessageP has a default handler for Send.sendDone, TestAMC does not have a default handler for Send.send. Otherwise, you could forget to wire TestAMC.Send and the program would compile fine but never send packets. Defaults should only be used when an interface is not necessary for the proper execution of a component. This almost always involves parameterized interfaces, as it's rare that all of the parameter values are used.

8.4 Attributes

Attributes are a way to associate metadata with program statements. nesC attributes are based on the approach taken with Java annotations [Chapter 9]. The full details are

8.4 Attributes

beyond the scope of this book, but it's worthwhile to present the most common attributes and how they are used. nesC attributes look like Java annotations. An attribute declaration is a struct declaration where the name is preceded by @. Attributes can therefore have fields, which can be initialized. We've already seen three attributes, @integer, @number, and @combine. For instance, the combine attribute we saw in Section 4.4.3 takes a string as argument, behaving as if[3] it were declared by:

```
struct @combine {
  char *function_name;
};

// Using the combine attribute in a type definition
typedef uint8_t error_t @combine("ecombine");
```

The declaration of error_t is annotated with the string "ecombine", specifying the name of the combine function for error_t. The argument in parentheses actually uses the same syntax (minus the {}) as a regular C initializer for the attribute's struct definition.

User-defined attributes can be attached to most names (components, interfaces, typedefs, variables, functions, etc) in nesC programs:

```
struct @atleastonce { };

configuration LedsC {
  provides interface Init @atleastonce();
  provides interface Leds;
}
```

Listing 8.21 nesC attributes

This example shows the declaration of the atleastonce attribute, which has no fields. The configuration LedsC annotates its Init interface with the attribute. By default, the @atleastonce attribute doesn't do anything. But the nesC compiler has tools to output information about an application, including attributes. Part of the default TinyOS compilation process includes running a script that checks wiring constraints, of which atleastonce is one (the others are atmostonce, exactlyonce). The tool checks that an interface annotated with atleastonce is wired to at least once. In the case of something like Init, the utility of this check is pretty clear: you can check at compile-time that your component is being initialized.

Attributes provide a way to extend the nesC language without introducing new keywords, which could break older code. The current common attributes are:

- **@spontaneous**: this function might be called from outside the nesC program, and so should not be pruned by nesC's dead code elimination. This attribute is needed for interrupt handlers and whenever you want to link binaries (e.g. with TOSSIM).
- **@C**: this function should be considered a C, rather than a nesC, function. Specifically, if a component defines a function with this attribute, then it is not made private to the

---
[3] combine is built in to the nesC compiler.

component. This attribute is needed for when C code needs to call nesC code (e.g. in TOSSIM).

- **@hwevent**: this function will be called as a result of a hardware interrupt. Implies spontaneous.
- **@atomic_hwevent**: this function will be called as a result of a hardware interrupt, and will execute in an atomic section. Implies spontaneous. The distinction between this attribute and @hwevent is needed so nesC can know whether additional atomic sections are needed.
- **@atmostonce**: this interface must be wired to zero or one times.
- **@atleastonce**: this interface must be wired one or more times.
- **@exactlyonce**: this interface must be wired to once, no more, no less.
- **@integer**: this type parameter to a generic component must be an integer. This attribute allows generic components to use integer operations (arithmetic, shifts, bit-operations) on the type, which is important for things like scaling timers.
- **@number**: this type parameter to a generic component must be an integer or floating-point type. This attribute allows generic components to use arithmetic operations on the type.
- **@combine**: this attribute is used to specify a combine function for a type when the type is declared. It takes a parameter, a string of the name of the combine function.

Earlier versions of nesC used gcc's attribute syntax to specify these attributes, however this is now deprecated. You may however still see it in some older code, e.g. in TinyOS 1.x:

```
module RealMain { ... }
implementation {
  int main() __attribute__ ((C, spontaneous)) {
  ...
```

8.5 Exercises

1. The standard QueueC implementation uses an array to store the queue. Write an alternative implementation that uses a linked list. The maximum size of the list should be a parameter to the generic component's constructor.
2. Extend your linked-list to provide an additional interface, List, which allows insertion, deletion, and peeking at any point in the list.
3. Write a stack data structure component that takes a max stack depth and data type as parameters.
4. Write an additional dispatch layer on top of AM, which provides a parameterized interface to a single AM id by adding a one-byte header.

9 Advanced wiring

Chapter 4 introduced the basics of nesC component wiring, explaining the three wiring operators and how multiple wirings work. Chapter 8 introduced parameterized interfaces, which enable a single component implementation to efficiently provide an interface to many clients. Much of the most intricate nesC programming involves using parameterized interfaces. This chapter focuses on generic configurations, and how they can use parameterized interfaces to provide robust and flexible abstractions.

9.1 unique() and uniqueCount()

Parameterized interfaces were originally intended to support abstractions like active messaging. It turns out, however, that they are much more powerful than that. If you look at the structure of most basic TinyOS 2.0 abstractions, there's a parameterized interface in there somewhere. The ability to specify compile-time constants outside of modules, combined with dispatch, means that we can use parameterized interfaces to distinguish between many different callers. A component can provide a service through a parameterized interface, and every client that needs to use the service can wire to a different parameter identifier. For split-phase calls, this means that you can avoid fan-out on the completion event. Consider these two examples:

```
components RouterC, SourceAC, SourceBC;
SourceAC.Send -> RouterC;
SourceBC.Send -> RouterC;
```

versus

```
components RouterC, SourceAC, SourceBC;
SourceAC.Send -> RouterC.Send[0];
SourceBC.Send -> RouterC.Send[1];
```

In both cases, SourceAC and SourceBC can call Send.send. In the first case, when RouterC signals Send.sendDone, that signal will fan-out to both SourceAC and SourceBC, who will have to determine – by examining the message pointer, or internal state variables – whether the event is intended for them or someone else. In the second case, however, if RouterC keeps the parameter identifier passed in the call to Send, then it can signal the appropriate completion event. For example, SourceBC calls Send.send, RouterC stores the identifier 1, and when it signals sendDone it signals it on Send.sendDone[1](...).

The combination of parameterized interfaces and generic configurations allows TinyOS 2.0 to virtualize abstractions at compile-time, using only as much memory as is needed and without using function pointers. Let's return to the timer example, where this abstraction is particularly powerful. The timer component HilTimerMilliC has the following signature:

```
configuration HilTimerMilliC {
  provides interface Timer<TMilli>[uint8_t];
}
```

Listing 9.1 Partial HilTimerMilliC signature

Because Timer is parameterized, many different components can wire to separate interface instances. When a component calls Timer.startPeriodic, nesC fills in the parameter identifier, which the timer implementation can use to keep track of which timer is being told to start. Similarly, the timer implementation can signal Timer.fired on specific timer instances.

For things such as network protocols, where the parameter to an interface is a basis for communication and interoperability, the actual parameter used is important. For example, if you have two different compilations of the same application, but one wires a protocol with

```
RouterC.Send -> ActiveMessageC.Send[210];
```

while the other wires it with

```
RouterC.Send -> ActiveMessageC.Send[211];
```

then they will not be able to communicate. In these cases, the parameter used is shared across nodes, and so needs to be globally consistent. Similarly, if you had two protocols wire to the same AM type, then this is a basic conflict that an application developer is going to have to resolve. Generally, protocols use named constants (enums) to avoid these kinds of typos.

9.1.1 unique

With timers and the Send client example above, however, there is no such restriction. The parameter represents a unique client identifier, rather than a piece of shared data. A client doesn't care which timer it wires to, as long as it wires to one that nobody else does. For this case, rather than force clients to guess identifiers and hope there is no collision, nesC provides a special compile-time function, unique().

It is a compile-time function because it is resolved at compile time. When nesC compiles an application, it replaces each call to unique() by an integer. The unique function takes a string key as an argument, and promises that every instance of the function with the same key will return a unique value. Two calls to unique with different keys can return the same value. So if two components, AppOneC and AppTwoC, both want timers, they could do this

```
AppOneC.Timer -> HilTimerMilliC.Timer[unique("Timer")];
AppTwoC.Timer -> HilTimerMilliC.Timer[unique("Timer")];
```

and be assured that they will have distinct timer instances.[1] If there are n calls to unique, then the unique values will be in the range of $0, 1, \ldots, n-1$.

9.1.2 uniqueCount

In these examples – Timer and Send – there is an additional factor to consider: each client requires the component to store some amount of state. For example, RouterC has to keep track of pending messages, and timer systems have to keep track of the period of each timer, how long until it fires, and whether it's active. Because the calls to unique define the set of valid client identifiers, nesC has a second compile-time function, uniqueCount(). This function also takes a string key. If there are n calls to unique with a given key (returning values $0, 1, \ldots, n-1$), then uniqueCount returns n. Like unique, uniqueCount is resolved at compile-time so can be used, e.g. to set array sizes.

Being able to count the number of unique clients allows a component to allocate the right amount of state to support them. Early versions of nesC didn't have the uniqueCount function: components were forced to allocate a fixed amount of state. If there were more clients than the state could support, one or more would fail at run-time. If there were fewer clients than the state could support, then there was wasted RAM. Because a component can count the number of clients and know the set of client identifiers that will be used, it can promise that each client will be able to work and use the minimum amount of RAM needed. Returning to the timer example from above:

```
AppOneC.Timer -> HilTimerMilliC.Timer[unique("Timer")];
AppTwoC.Timer -> HilTimerMilliC.Timer[unique("Timer")];
```

and HilTimerMilliC could allocate state for each client:

```
timer_t timers[uniqueCount("Timer")];
```

Assuming the above two were the only timers, then HilTimerMilliC would allocate two timer structures. If we assume that AppOneC.Timer was assigned identifier 0 and AppTwoC.Timer was assigned identifier 1, then HilTimerMilliC can directly use the parameter as an index into the state array.

9.1.3 Example: HilTimerMilliC and VirtualizeTimerC

This isn't how HilTimerMilliC works: it's actually a bit more complicated, as it is a configuration built out of a set of generic components (see Section 9.3). As hardware platforms differ in their timer hardware support, HilTimerMilliC is a platform-specific

---

[1] In practice, clients rarely call unique() directly. Instead, these calls are encapsulated inside generic configurations, as we will see below. One common problem with unique() encountered in TinyOS 1.x is that a mistyped key will generate a non-unique value and possibly cause very strange behavior.

Advanced wiring

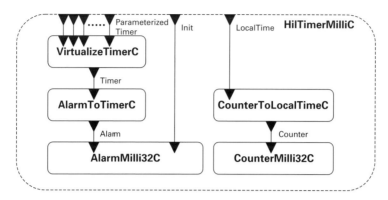

Figure 9.1 Partial component structure of a typical HilTimerMilliC.

component. Typically (Figure 9.1), it builds up a single timer from hardware counters and interrupts, then virtualizes this single timer into many timers. The virtualizing component, VirtualizeTimerC, stores state for each of the timers it provides, and schedules the single timer it uses to fire when the next provider needs to fire. VirtualizeTimerC is a generic module:

```
generic module VirtualizeTimerC(typedef precision_tag, int max_timers) {
  provides interface Timer<precision_tag> as Timer[uint8_t num];
  uses interface Timer<precision_tag> as TimerFrom;
}
```

Listing 9.2 VirtualizeTimerC

HilTimerMilliC must pass the number of timers in the system, obtained using uniqueCount(), to the new VirtualizeTimerC instance:

```
configuration HilTimerMilliC {
  provides interface Init;
  provides interface Timer<TMilli> as TimerMilli[uint8_t num];
  provides interface LocalTime<TMilli>;
}
implementation {
  enum {
    TIMER_COUNT = uniqueCount(UQ_TIMER_MILLI)
  };

  components new VirtualizeTimerC(TMilli, TIMER_COUNT);
  TimerMilli = VirtualizeTimerC;x
  // More code for wiring VirtualizeTimerC.TimerFrom
}
```

Listing 9.3 Instantiating VirtualizeTimerC

9.1 unique() and uniqueCount()

Rather than directly pass a string to uniqueCount(), HilTimerMilliC passes UQ_TIMER_MILLI, which is a #define (from `Timer.h`) for the string "HilTimerMilliC.Timer". This is a common approach in TinyOS. Using a #define makes it harder to run into bugs caused by errors in the string: chances are that a typo in the uses of the #define will be a compile-time error. This is generally a good practice for components that depend on unique strings.

> **Programming Hint 19** IF A COMPONENT DEPENDS ON UNIQUE, THEN #DEFINE THE STRING TO USE IN A HEADER FILE, TO PREVENT BUGS DUE TO STRING TYPOS.

VirtualizeTimerC, in turn, allocates an array of timer state structures, whose size is determined by its max_timers parameter:

```
enum {
  NUM_TIMERS = max_timers,
  END_OF_LIST = 255,
};
typedef struct {
  // details elided
} Timer_t;

Timer_t m_timers[NUM_TIMERS];
```

Listing 9.4 VirtualizeTimerC state allocation

When a client calls VirtualizeTimerC's Timer interface (through HilTimerMilliC), the parameter indicates which client is making the call. VirtualizeTimerC uses the client ID as an index into the state array. For example:

```
void startTimer(uint8_t num, uint32_t t0, uint32_t dt, bool isoneshot) {
  Timer_t* timer = &m_timers[num];
  // update timer structure
}

command void Timer.startPeriodic[uint8_t num](uint32_t dt) {
  startTimer(num, call TimerFrom.getNow(), dt, FALSE);
}
```

This software structure assumes that each component that needs a timer wires to HilTimerMilliC's parameterized TimerMilli using a call to unique, passing UQ_TIMER_MILLI as the key:

```
SomeComponentC.Timer -> HilTimerMilliC.TimerMilli[unique(UQ_TIMER_MILLI)];
```

Requiring a programmer to know a service's key is problematic and bug-prone. While a #define for the key reduces programming errors compared to actual strings, it's still very easy for a programmer to accidentally use the wrong key, especially if an application uses many services. Finding such a bug is a nightmare. This is essentially a problem of code duplication: even though every Timer user wants to wire in the same way, each one

150 Advanced wiring

has to repeat a particular code sequence. Generic configurations solve this problem: in practice, services don't require users to wire to parameterized interfaces.

9.2 Generic configurations

Generic modules are a way to reuse code and separate common abstractions into well-tested building blocks (there only needs to be one FIFO queue implementation, for example). nesC also has generic configurations, which are a very powerful tool for building TinyOS abstractions and services. However, just as configurations are harder for a novice programmer to understand than modules, generic configurations are a bit more challenging than generic modules.

The best way to describe what role a generic configuration can play in a software design is to start from first principles.

A module is a component that contains executable code; a configuration defines relationships between components to form a higher-level abstraction; a generic module is a reusable piece of executable code; therefore, a generic configuration is a reusable set of relationships that form a higher-level abstraction.

Several examples in this book have mentioned and described HilTimerMilliC. But if you look at TinyOS code, there is only one component that references it. Although it is a very important component, programs never directly name it. It is the core part of the timer service, but applications that need timers instantiate a generic component named TimerMilliC.

Before delving into generic configurations, however, let's consider what code looks like without them. Let's say we have HilTimerMilliC, and nothing more. Many components need timers; HilTimerMilliC enables this through its parameterized interface. Remember that HilTimerMilliC encapsulates an instance of VirtualizeTimerC, whose size parameter is a call to unique(UQ_TIMER_MILLI). This means that if a component AppP needs a timer, then its configuration AppC must wire it like this:

```
enum {
  TIMER_KEY = unique(UQ_TIMER_MILLI)
};
AppP.Timer -> HilTimerMilliC.TimerMilli[TIMER_KEY];
```

Note in passing that this snippet (and the earlier definition of HilTimerMilliC) take advantage of the fact that the body of configurations can contain type and constant declarations, whose scope extends to the end of the configuration's implementation section. Providing a name like TIMER_KEY for a unique value is one of the most common ways this facility is used.

9.2.1 TimerMilliC

TimerMilliC is TinyOS's standard standard millisecond timer abstraction, designed to simplify usage of TinyOS's virtualized timer service, HilTimerMilliC. TimerMilliC is a generic configuration that provides a single Timer interface. Its implementation wires this interface to an instance of the underlying parameterized Timer interface using the

9.2 Generic configurations

right unique key. This means that unique() is called in only one file; as long as all components allocate timers with TimerMilliC, there is no chance of a key match mistake. TimerMilliC's implementation is very simple:

```
generic configuration TimerMilliC() {
  provides interface Timer<TMilli>;
}
implementation {
  components TimerMilliP;
  Timer = TimerMilliP.TimerMilli[unique(UQ_TIMER_MILLI)];
}
```

Listing 9.5 The TimerMilliC generic configuration

TimerMilliP is a singleton configuration that auto-wires HilTimerMilliC to the boot sequence and exports HilTimerMilliC's parameterized interface:

```
configuration TimerMilliP {
  provides interface Timer<TMilli> as TimerMilli[uint8_t id];
}
implementation {
  components HilTimerMilliC, MainC;
  MainC.SoftwareInit -> HilTimerMilliC;
  TimerMilli = HilTimerMilliC;
}
```

Listing 9.6 TimerMilliP auto-wires HilTimerMilliC to Main.SoftwareInit

TimerMilliC encapsulates a wiring pattern – wiring to the timer service with a call to unique – for other components to use. When a component instantiates a TimerMilliC, it creates a copy of the TimerMilliC code, which includes a call to unique(UQ_TIMER_MILLI). The code

```
components X, new TimerMilliC();
X.Timer -> TimerMilliC;
```

is essentially identical to:

```
components X, TimerMilliP;
X.Timer -> TimerMilliP.TimerMilli[unique(UQ_TIMER_MILLI)];
```

Let's step through the complete wiring path for an application that creates a timer. BlinkAppC wires the BlinkC module to its three timers:

```
configuration BlinkAppC{}
implementation {
  components MainC, BlinkC, LedsC;
  components new TimerMilliC() as Timer0;
  components new TimerMilliC() as Timer1;
  components new TimerMilliC() as Timer2;
```

Advanced wiring

```
    BlinkC -> MainC.Boot;
    MainC.SoftwareInit -> LedsC;

    BlinkC.Timer0 -> Timer0;
    BlinkC.Timer1 -> Timer1;
    BlinkC.Timer2 -> Timer2;
    BlinkC.Leds -> LedsC;
}
```

Listing 9.7 The Blink application

The HilTimerMilliC configuration wires its Timer interface to a VirtualizeTimerC timer virtualization component (Section 9.1.3). Thus, wiring BlinkC.Timer0 to Timer0 establishes this wiring chain:

```
BlinkC.Timer0 -> Timer0.Timer
enum { K = unique(UQ_TIMER_MILLI) };
Timer0.Timer = TimerMilliP.TimerMilli[K]
TimerMilliP.TimerMilli[K] = HilTimerMilliC[K]
HilTimerMilliC[K] = VirtualizeTimerC.Timer[K]
```

Listing 9.8 The full module-to-module wiring chain in Blink (BlinkC to VirtualizeTimerC)

BlinkC and VirtualizeTimerC are the two modules; the intervening components are all configurations. When nesC compiles this code, all of the intermediate layers will be stripped away, and BlinkC.Timer0.start will be a direct function call on VirtualizeTimerC.Timer[...].start.

Many of TinyOS's basic services use this pattern of a generic configuration managing a *keyspace* (a set of identifiers, see Section 10.3) for a parameterized interface. Sometimes the mapping is more complex than TimerMilliC, as we will see below when we describe BlockStorageC, one of TinyOS's non-volatile-storage abstractions.

9.2.2 CC2420SpiC

Another, more complex example of using generic configurations is CC2420SpiC. This component provides access to the CC2420 radio over an SPI bus. When the radio stack software wants to interact with the radio, it makes calls on an instance of this component. For example, telling the CC2420 to send a packet if there is a clear channel involves writing to one of the radio's registers (TXONCCA). To write to the register, the stack sends a small series of bytes over the bus, which basically say "I'm writing to register number X with value Y." The very fast speed of the bus means that small operations such as these can made synchronous without any significant concurrency problems.

In addition to small register reads and writes, the chip also supports accessing the receive and transmit buffers, which are 128-byte regions of memory, as well as

9.2 Generic configurations

the radio's configuration memory, which stores things such as cryptographic keys and the local address (which is used for determining whether to send an acknowledgement). These operations may take a while, so are split-phase. For example, before the stack writes to TXONCCA to send a packet, it must first execute a split-phase write of the packet contents with the CC2420Fifo interface (the receive and transmit buffers are FIFO memories).

All of the operations boil down to four interfaces:

- **CC2420Strobe**: Access to a command register. Writing a command register tells the radio to take an action, such as transmit a packet, clear its packet buffers, or transition to transmit mode. This interface has a single command, strobe, which writes to the register.
- **CC2420Register**: Access to a data register. These registers can be both read and written, and store things such as hardware configuration, addressing mode, and clear channel assessment thresholds. This interface supports reads and writes as single-phase operations.
- **CC2420Ram**: Access to configuration memory. This interface supports both reads and writes, as split-phase operations.
- **CC2420Fifo**: Access to the receive and transmit FIFO memory buffers. This interface supports both reads and writes, as split-phase operations. While one can write to the receive buffer, the CC2420 supports this only for debugging purposes.

A component that needs to interact with the CC2420 instantiates an instance of CC2420SpiC:

```
generic configuration CC2420SpiC() {

  provides interface Resource;

  provides interface CC2420Strobe as SFLUSHRX;
  provides interface CC2420Strobe as SFLUSHTX;
  provides interface CC2420Strobe as SNOP;
  provides interface CC2420Strobe as SRXON;
  provides interface CC2420Strobe as SRFOFF;
  provides interface CC2420Strobe as STXON;
  provides interface CC2420Strobe as STXONCCA;
  provides interface CC2420Strobe as SXOSCON;
  provides interface CC2420Strobe as SXOSCOFF;

  provides interface CC2420Register as FSCTRL;
  provides interface CC2420Register as IOCFG0;
  provides interface CC2420Register as IOCFG1;
  provides interface CC2420Register as MDMCTRL0;
  provides interface CC2420Register as MDMCTRL1;
  provides interface CC2420Register as TXCTRL;

  provides interface CC2420Ram as IEEEADR;
  provides interface CC2420Ram as PANID;
```

Advanced wiring

```
  provides interface CC2420Ram as SHORTADR;
  provides interface CC2420Ram as TXFIFO_RAM;

  provides interface CC2420Fifo as RXFIFO;
  provides interface CC2420Fifo as TXFIFO;
}
```

Listing 9.9 CC2420SpiC

CC2420SpiC takes the implementation of the SPI protocol (CC2420SpiP) and wires it to the platform's raw SPI implementation. The raw SPI implementation has two interfaces: SpiByte, for writing a byte as a single-phase operation, and SpiPacket, for writing a series of bytes as a split-phase operation. The SPI protocol is bidirectional. To read bytes from the chip, the stack has to write onto the bus. The chip also writes onto the bus, but it is clocked by the CPU's writes. The write operation therefore takes a uint8_t as the byte to write, and return a uint8_t byte representing the reply.

The protocol implementation uses an interesting approach to have a simple implementation that can also find compile-time wiring errors. While CC2420SpiC provides each register as a separate interface, the executable logic (CC2420SpiP) provides a parameterized interface:

```
configuration CC2420SpiP {
  provides interface CC2420Fifo as Fifo[ uint8_t id ];
  provides interface CC2420Ram as Ram[ uint16_t id ];
  provides interface CC2420Register as Reg[ uint8_t id ];
  provides interface CC2420Strobe as Strobe[ uint8_t id ];
}
```

Listing 9.10 CC2420SpiP

Each CC2420 register has a unique identifier, which is a small integer. Having a separate implementation for each register operation wastes code space and the code repetition would be an easy way to introduce bugs. So CC2420SpiP has a single implementation, which takes a compile-time parameter, the register identifier. However, not all values of a uint8_t are valid registers, so allowing components to wire directly to the parameterized interface could lead to invalid wirings. Of course, CC2420SpiP could incorporate some run-time checks to make sure that register values are valid, but this wastes CPU cycles, especially when the parameters should always be valid. So CC2420SpiC maps a subset of the valid parameters into interface instances. It only maps a subset because there are some debugging registers the stack doesn't need to use. The implementation looks like this:

```
configuration CC2420SpiC { ...}
implementation {
  ...
  components CC2420SpiP as Spi;
```

9.2 Generic configurations

```
    SFLUSHRX = Spi.Strobe[CC2420_SFLUSHRX];
    SFLUSHTX = Spi.Strobe[CC2420_SFLUSHTX];
    SNOP = Spi.Strobe[CC2420_SNOP];
    SRXON = Spi.Strobe[CC2420_SRXON];
    SRFOFF = Spi.Strobe[CC2420_SRFOFF];
    STXON = Spi.Strobe[CC2420_STXON];
    STXONCCA = Spi.Strobe[CC2420_STXONCCA];
    SXOSCON = Spi.Strobe[CC2420_SXOSCON];
    SXOSCOFF = Spi.Strobe[CC2420_SXOSCOFF];
}
```

Listing 9.11 CC2420SpiC mappings to CC2420SpiP

This approach gives us the best of both worlds: there is a single function for writing to a strobe register, which takes as an argument which register to write to, and the argument does not need run-time checking. However, the caller does not have to provide this identifier, and so cannot by accident specify an invalid one. Components that wire to CC2420SpiC can only wire to valid strobe registers, and rather than doing

```
call CC2420Strobe.strobe(CC2420_STXONCCA);
```

they write

```
call TXONCCA.strobe();
```

Let's look at what this means at a function level. As we saw above, every component which wires to TXONCCA on an instance of CC2420SpiC wires to Spi.Strobe[CC242_TXONCCA]. This wiring terminates in the CC2420SpiP module:

```
async command cc2420_status_t Strobe.strobe[ uint8_t addr ]() {
    return call SpiByte.write(addr);
}
```

Listing 9.12 The strobe implementation

which writes a single byte to the bus and returns the status result. To step through each layer,

1. A component (e.g. CC2420TransmitP) calls TXONCCA.strobe() on an instance of CC2420SpiC
2. The nesC wiring transforms this call into CC2420SpiP.Strobe[CC2420_TXONCCA].strobe().

After optimizations and inlining, the original statement `call TXONCCA.strobe()` effectively becomes

```
call SpiByte.write(CC2420_TXONCCA);
```

with possible further optimization down into the SPI layer (it might just inline SpiByte.write into the function, removing any need for function calls).

If a function has an argument which is one of a small number of constants, consider defining it as a few separate functions to prevent bugs. If the functions of an interface all have an argument that's almost always a constant within a large range, consider using a parameterized interface to save code space. If the functions of an interface all have an argument that's a constant within a large range but with only certain valid values, implement it as a parameterized interface but expose it as individual interfaces, to both minimize code size and prevent bugs.

9.2.3 AMSenderC

HilTimerMilliC virtualizes a hardware timer and makes independent instances available via a parameterized interface. TimerMilliC is thus reasonably simple: all it really does is encapsulate a wiring with unique() to prevent client collisions and to simplify wiring.

Active messages are slightly different. The basic platform-supplied active message component, ActiveMessageC, provides AMSend, parameterized by the AM identifier. However, ActiveMessageC can only have a single packet outstanding at any time. If it is already sending a packet and a component calls AMSend.send, ActiveMessageC returns FAIL or EBUSY. From the perspective of a caller, this is a bit of a pain. If it wants to send the packet, it has to wait until the radio is free, but doesn't have a very easy way of figuring out when this will occur.

TinyOS 1.x had a global (not parameterized) sendDone event, which the radio would signal whenever it finished sending any packet. That way, if a component tried to send and received a FAIL, it could try to resend when it handled the global sendDone event. This mostly works, except that if multiple components wire to sendDone, then the fan-out determines the priority of the send requests. For example, if a hog of a component handles sendDone and happens to be first in the fan-out, it will always be the first to call Send.send and will monopolize the radio.

TinyOS 2.0 solves this problem through the AMSenderC component, which is a generic configuration. AMSenderC is a *virtualized* abstraction: every instance of AMSenderC acts like ActiveMessageC. That is, each AMSenderC can handle a single outgoing packet. This means that each component that wires to an AMSenderC can act independently of the other components, and not worry about fan-out scheduling. The one-deep queue of ActiveMessageC is replaced by N one-deep queues, one for each of the N clients.

A one-deep queue per AMSenderC isn't sufficient. There's also the question of what order the senders get to send their packets. Under the covers, what the active message layer does is maintain an array of N pending packets, where N is the number of AMSenderC components. Each AMSenderC is a client of the active message sending abstraction, and so has a client identifier that indexes into this array. The implementation keeps track of the last client that was able to send a packet, and makes sure that everyone else waiting gets a chance before that client does again (this is known as round-robin scheduling).

9.2 Generic configurations

Accomplishing this is a little trickier than TimerMilliC, because a request to send has a few parameters. With Timer, those parameters (period, single-shot vs. repeat) are state that the timer implementation has to keep track of in the first place. With AMSenderC, it's a bit different: those parameters just need to be stored until the call to the underlying ActiveMessageC. The send queue could just store all of these parameters, that uses up four extra bytes of RAM per entry (two for the destination, one for the AM type, and one for the length).

It turns out that the Packet and AMPacket interfaces have operations exactly for this situation. They allow a component to get and set packet fields. For example, a component can call Packet.setLength to set the length field and recover it with Packet.length. Components that just need basic send or receive abstractions can just use AMSend or Receive. The Packet interface, though, allows data structures such as queues to store temporary state within the packet and then recover it when it's time to actually send so it can be passed as parameters. This means that the AM send queue with n clients allocates a total of $(2n + 1)$ bytes of state, as pointers on microcontrollers are usually two bytes (on the intelmote2, though, they're four bytes, so it allocates $4n + 1$).

This means that the AMSenderC abstraction needs to do the following things:

1. Provide an AMSend interface
2. Store the AMSend.send parameters before putting a packet on the queue
3. Statically allocate a single private queue entry
4. Store a send request packet in the queue entry when its entry is not occupied
5. When it's actually time to send the packet, reconstitute the send parameters and call ActiveMessageC

AMSenderC has one additional complication: keyspaces (sets of identifiers). ActiveMessageC provides AMSend based on the AM type keyspace while the send queue has a client identifier keyspace for keeping track of which AMSenderC is sending. Because the queue needs to be able to send any AM type, it uses a parameterized AMSend and directly wires to ActiveMessageC.AMSend. Figure 9.2

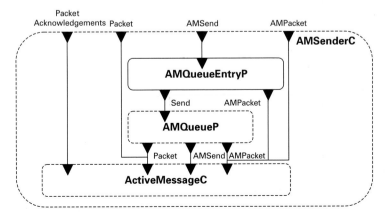

Figure 9.2 AMSenderC components

shows the resulting components: AMQueueEntryP implements the per-client logic needed to queue packets for a particular AM identifier, AMQueueImplP implements the queue, and AMQueueP wires the queue to ActiveMessageC. At run-time, a typical execution is:

1. Client calls AMSenderC's AMSend.send.
2. This calls end in module AMQueueEntryP, which stores the length, AM identifier, and destination in the packet.
3. AMQueueEntryP is a client of AMQueueImplP and calls Send.send with its client identifier.
4. AMQueueImplP checks that the client's queue entry is free and puts the packet into it.
5. Some time later, AMQueueImplP pulls the packet off the queue and calls AMSend.send on ActiveMessageC with the parameters that AMQueueEntryP stored.
6. When ActiveMessageC signals AMSend.sendDone, AMQueueImplP signals Send.sendDone to AMQueueEntryP, which signals AMSend.sendDone to the original calling component.

This is the code for AMSenderC:

```
generic configuration AMSenderC(am_id_t AMId) {
  provides {
    interface AMSend;
    interface Packet;
    interface AMPacket;
    interface PacketAcknowledgements as Acks;
  }
}

implementation {
  components new AMQueueEntryP(AMId) as AMQueueEntryP;
  components AMQueueP, ActiveMessageC;

  AMQueueEntryP.Send -> AMQueueP.Send[unique(UQ_AMQUEUE_SEND)];
  AMQueueEntryP.AMPacket -> ActiveMessageC;

  AMSend   = AMQueueEntryP;
  Packet   = ActiveMessageC;
  AMPacket = ActiveMessageC;
  Acks     = ActiveMessageC;
}
```

Listing 9.13 The AMSenderC generic configuration

A send queue entry is responsible for storing send information in a packet:

```
generic module AMQueueEntryP(am_id_t amId) {
  provides interface AMSend;
  uses{
    interface Send;
    interface AMPacket;
  }
}

implementation {

  command error_t AMSend.send(am_addr_t dest,
                              message_t* msg,
                              uint8_t len) {
    call AMPacket.setDestination(msg, dest);
    call AMPacket.setType(msg, amId);
    return call Send.send(msg, len);
  }

  command error_t AMSend.cancel(message_t* msg) {
    return call Send.cancel(msg);
  }

  event void Send.sendDone(message_t* m, error_t err) {
    signal AMSend.sendDone(m, err);
  }

  command uint8_t AMSend.maxPayloadLength() {
    return call Send.maxPayloadLength();
  }

  command void* AMSend.getPayload(message_t* m) {
    return call Send.getPayload(m);
  }

}
```

Listing 9.14 AMSendQueueEntryP

The queue itself sits on top of ActiveMessageC:

```
configuration AMQueueP {
  provides interface Send[uint8_t client];
}

implementation {
  components AMQueueImplP, ActiveMessageC;

  Send = AMQueueImplP;
```

```
    AMQueueImplP.AMSend   -> ActiveMessageC;
    AMQueueImplP.AMPacket -> ActiveMessageC;
    AMQueueImplP.Packet   -> ActiveMessageC;

}
```

Listing 9.15 AMQueueP

Finally, within AMSendQueueImplP, the logic to send a packet looks like this:

```
nextPacket();
if (current == QUEUE_EMPTY) {
  return;
}
else {
  message_t* msg;
  am_id_t id;
  am_addr_t addr;
  uint8_t len;
  msg = queue[current];
  id = call AMPacket.getType(msg);
  addr = call AMPacket.getDestination(msg);
  len = call Packet.getLength(msg);
  if (call AMSend.send[id](addr, msg, len) == SUCCESS) {
    ...
  ...
}
```

Listing 9.16 AMSendQueueImplP pseudocode

9.2.4 BlockStorageC

One of the most difficult parts of nesC programming is using parameterized interfaces and managing their keyspaces. As we saw in Section 6.5, BlockStorageC is one of TinyOS's three storage abstractions, and provides random read/write access to a storage volume (a contiguous area on a flash chip). BlockStorageC is a good example of using generics, because it is non-trivial use of wiring to build an abstraction from underlying components. BlockStorageC deals with four different sets of parameterized interfaces.

Every volume has a unique identifier, and every BlockStorageC is associated with a single volume. However, there can be multiple BlockStorageC components accessing multiple volumes, and not all volumes may have BlockStorageC components (they may be unused or accessed by a different abstraction). A client has to be associated with a volume so that the underlying code can generate an absolute offset into the chip from a relative offset within a volume. For example, if a 1 MB flash chip is divided into two 512 kB volumes, then address 16k on volume 1 is address 528k on the chip. This means that there are at least two keyspaces. The first keyspace is the volume ID keyspace, and the second is the client ID keyspace.

In practice, there is a third keyspace, which is used to arbitrate access amongst all clients to the flash chip (see Section 11.2). This keyspace is shared between block clients, logging clients, and other abstractions that need exclusive access to the flash chip. So, all in all, BlockStorageC has to manage three different keyspaces:

1. Client key: which block storage client this is (for block storage client state)
2. Chip key: which client to the flash chip this is (for arbitration of the shared resource)
3. Volume key: which volume this client accesses (for calculating absolute offsets in the chip)

Both the client key and chip key are generated with unique(). The client key is only among BlockStorageC components, so it uses a string UQ_BLOCK_STORAGE defined in BlockStorage.h. The chip key is shared across all components that use underlying chip. In the case of the AT45DB chip (used in the micaz platform), the string is UQ_AT45DB defined in At45db.h. The volume key (VOLUME_XXX) is not generated by unique, as it is generated from the XML file that specifies the flash chip layout (Figure 6.3, page 102).

After all of that introduction, you might think that BlockStorageC is many lines of code. It isn't: it only has four wiring statements, which we'll step through one by one:

```
generic configuration BlockStorageC(volume_id_t volid) {
  provides {
    interface BlockWrite;
    interface BlockRead;
  }
}
implementation {
  enum {
    BLOCK_ID = unique(UQ_BLOCK_STORAGE),
    RESOURCE_ID = unique(UQ_AT45DB)
  };

  components BlockStorageP, WireBlockStorageP, StorageManagerC, At45dbC;

  BlockWrite = BlockStorageP.BlockWrite[BLOCK_ID];
  BlockRead = BlockStorageP.BlockRead[BLOCK_ID];

  BlockStorageP.At45dbVolume[BLOCK_ID] -> StorageManagerC.At45dbVolume[volid];
  BlockStorageP.Resource[BLOCK_ID] -> At45dbC.Resource[RESOURCE_ID];
}
```

Listing 9.17 BlockStorageC

The first two lines,

```
BlockWrite = BlockStorageP.BlockWrite[BLOCK_ID];
BlockRead = BlockStorageP.BlockRead[BLOCK_ID];
```

make the BlockWrite and BlockRead interfaces clients of the service that implements them, BlockStorageP. When a component wired to a BlockStorageC calls to read or write from a block, nesC automatically includes a client ID into the call by the time it reaches the implementation.

The next line

```
BlockStorageP.At45dbVolume[BLOCK_ID] -> StorageManagerC.At45dbVolume[volid];
```

translates between the client and volume keyspaces. When BlockStorageP makes a call on the StorageManagerC, it includes the client ID in the call as an outgoing parameter. This client ID is bound to a volume ID. nesC automatically builds a switch statement that translates between the two, so that when StorageManagerC receives the call, nesC has filled in the volume ID as the parameter.

The final line,

```
BlockStorageP.Resource[BLOCK_ID] -> At45dbC.Resource[RESOURCE_ID];
```

is what allows the block storage client to cooperate with other clients (blocking and logging) for access to the actual flash chip. BlockStorageP makes each of its clients a client of the flash chip resource manager (Chapter 11.2).

Overall, the logic goes like this:

1. A component accesses volume V through an instance of BlockStorageC with client ID C
2. The component calls BlockStorageC to read from a block
3. It becomes a call on BlockStorageP with with parameter C
4. BlockStorageP notes that there is a call pending, stores the arguments in the state allocated for C, and requests the Resource with C, which maps to resource ID R
5. BlockStorageP receives the resource for client R (which maps back to C)
6. BlockStorageP requests operations on StorageManagerC with C, which maps to volume V

The two complicated parts are the mapping between key spaces. In the case of the client ID and resource ID, the keyspaces are used to distinguish different callers, especially for storing state. The volume keyspace, however, is a little different. It is used to calculate an offset into the storage medium. The motivation for its being a parameter of a parameterized interface is a bit different. It is more like AMSend, where the value is a constant and can be easily decoupled from the implementation. Rather than passing a volume ID into a module and forcing it to include the constant as an argument to every function call, putting it into a configuration lets nesC automatically generate code to include the constant in all calls and a dispatch table for all events.

9.3 Reusable component libraries

Implementing a solid and efficient timer subsystem is very difficult. TinyOS makes the task simpler by having a library of reusable components (tos/lib/timer) that provide many of the needed pieces of functionality. This library is a good example of

9.3 Reusable component libraries

a set of generic components which can be assembled to build useful abstractions: each supported microcontroller provides a timer system by layering these library components on top of some low-level hardware abstractions. Because many microcontrollers have several clock sources, most of these library components are generic components, so that a platform can readily provide several timer systems of different fidelities.

For example, this is the full code for HilTimerMilliC on the micaz platform – it is similar, but not identical to the typical structure shown in Figure 9.1, page 148. It is defined in `tos/platforms/mica`, which contains common abstractions across the entire mica family (e.g. mica2, mica2dot, micaz):

```
#include "Timer.h"

configuration HilTimerMilliC {
  provides interface Init;
  provides interface Timer<TMilli> as TimerMilli[uint8_t num];
  provides interface LocalTime<TMilli>;
}
implementation {

  enum {
    TIMER_COUNT = uniqueCount(UQ_TIMER_MILLI)
  };

  components AlarmCounterMilliP, new AlarmToTimerC(TMilli),
    new VirtualizeTimerC(TMilli, TIMER_COUNT),
    new CounterToLocalTimeC(TMilli);

  Init = AlarmCounterMilliP;

  TimerMilli = VirtualizeTimerC;
  VirtualizeTimerC.TimerFrom -> AlarmToTimerC;
  AlarmToTimerC.Alarm -> AlarmCounterMilliP;

  LocalTime = CounterToLocalTimeC;
  CounterToLocalTimeC.Counter -> AlarmCounterMilliP;
}
```

Listing 9.18 The full code of HilTimerMilliC

The only singleton component in this configuration is AlarmCounterMilliP, which is an abstraction of a low-level microcontroller timer. HilTimerMilliC uses three generic components on top of AlarmCounterMilliP to provide a full timer system:

- CounterToLocalTimerC turns a hardware counter into a local timebase;
- AlarmToTimerC turns an Alarm interface, which provides an interrupt-driven one-shot timer, into a Timer interface, which provides a synchronous timer with greater functionality;
- VirtualizeTimerC virtualizes a single Timer into *n* Timers, where *n* is a component argument.

All of the generics have the type TMilli as one of their arguments. These type arguments make sure that timer fidelities are not accidentally changed. For instance, VirtualizeTimerC

```
generic module VirtualizeTimerC(typedef precision_tag, int max_timers)
{
  provides interface Timer<precision_tag> as Timer[uint8_t num];
  uses interface Timer<precision_tag> as TimerFrom;
}
...
```

Listing 9.19 VirtualizeTimerC virtualizes a single timer

takes a single timer of fidelity precision_tag and virtualizes it into timer_count timers of precision_tag. The timer library also has components that translate between precisions.

HilTimerMilliC takes an interrupt-driven hardware timer – AlarmCounterMilliP – and turns it into a virtualized timer. It does this with three steps. The first step turns the interrupt-driven Alarm into a task-based Timer, with the generic component AlarmToTimerC:

```
AlarmToTimerC.Alarm -> AlarmCounterMilliP;
```

The second step virtualizes a single timer into many timers:

```
VirtualizeTimerC.TimerFrom -> AlarmToTimerC;
```

HilTimerMilliC then exports the parameterized timer interface:

```
TimerMilli = VirtualizeTimerC;
```

Additionally, some aspects of the timer system require being able to access a time base, for example, to specify when in the future a timer fires. So HilTimerMilliC takes a hardware counter and turns it into a local time component,

```
CounterToLocalTimeC.Counter -> AlarmCounterMilliP;
```

then exports the interface:

```
LocalTime = CounterToLocalTimeC;
```

Many of the components in the timer library are generics because a platform might need to provide a wide range of timers. For example, depending on the number of counters, compare registers, and their width, a platform might provide millisecond, microsecond, and 32 kHz timers. The variants of the MSP430 chip family that some platforms use, for example, can easily provide 32-bit millisecond and 32 kHz timers with a very low interrupt load.

Generic modules work very well for abstractions that have to allocate per-client state, such as timers. A generic module allows you to specify the size – the number of clients – in the configuration that instantiates the module, rather than within the module itself. For example, if VirtualizeTimerC were not a generic, then inside its code there would have to be a uniqueCount() with the proper key.

Programming Hint 20 WHENEVER WRITING A MODULE, CONSIDER MAKING IT MORE GENERAL-PURPOSE AND GENERIC. IN MOST CASES, MODULES MUST BE WRAPPED BY CONFIGURATIONS TO BE USEFUL, SO SINGLETON MODULES HAVE FEW ADVANTAGES.

9.4 Exercises

1. Use the TinyOS timer library to create a second-granularity timer abstraction, TimerSecondC. If you virtualize on top of a TimerMilliC, then the low-level timer only needs to scan one timer entry for all of the second-granularity timers. Be careful with the unique key to your HilTimerSecondC's parameterized interface.
2. The standard TinyOS collection layer, CTP (found in `tos/lib/net/ctp`), routes packets across multiple hops to a data sink node. CTP has a send queue of depth $C + F$, where C is the number of sending clients and F is the size of its pool of packets for forwarding. This means that a client can get up to $\frac{1}{C+F}$ of the available link throughput. Change CtpForwardingEngineP and CollectionSenderP so that there is a separate client queue, which sits on top of a single entry in CtpForwardingEngineP's queue, causing a client to get at most $\frac{1}{C \cdot F}$ of the link throughput.
3. Write a barrier for an arbitrary number of clients. A barrier is a concurrency primitive that blocks until all threads reach it. In TinyOS, since there are no threads, this means block until all callers reach it. Write a split-phase interface, `Barrier`, which has a single command, `wait` and a single event `pass`. When all clients have called wait, the barrier should signal pass to all of them.

10 Design patterns

To quote the Gang of Four, design patterns are "descriptions of communicating objects and classes that are customized to solve a general design problem in a particular context." [3] In the components we've seen so far, we see several recurring patterns, such as the use of parameterized interfaces to implement services with multiple clients (VirtualizeTimerC, Section 9.1.3), or one component wrapping another (RandomC, Section 4.2). In this chapter, in the spirit of the Gang of Four's original design patterns work, we attempt to formalize a number of these patterns, based on our observations during TinyOS's development.

This chapter presents eight nesC design patterns: three behavioral (relating to component interaction): Dispatcher, Decorator, and Adapter, three structural (relating to how applications are structured): Service Instance, Placeholder, and Facade and two namespace (management of identifiers such as message types): Keyspace and Keymap. Each pattern's presentation follows the model of the Design Patterns book. Each one has an *Intent*, which briefly describes its purpose. A more in-depth *Motivation* follows, providing an example drawn from TinyOS. *Applicable When* provides a succinct list of conditions for use and a component diagram shows the *Structure* of how components in the pattern interact.[1] In addition to our usual conventions for component diagrams, we attach folded sub-boxes to components to show relevant code snippets (a floating folded box represents source code in some other, unnamed, component). The diagram is followed by a *Participants* list explaining the role of each component. *Sample Code* shows an example nesC implementation, and *Known Uses* points to some uses of the pattern in TinyOS. *Consequences* describes how the pattern achieves its goals, and notes issues to consider when using it. Finally, *Related Patterns* compares to other relevant patterns.

10.1 Behavioral: Dispatcher

Intent
Dynamically select between a set of operations based on an identifier. Provides a way to easily extend or modify a system by adding or changing operations.

---

[1] This diagram is omitted for the Keyspace pattern as it is not concerned with component interactions.

Motivation

At a high level, sensor network applications execute operations in response to environmental input such as sensor readings or network packets. The operation's details are not important to the component that presents the input. We need to be able to easily extend and modify what inputs an application cares about, as well as the operation associated with an input.

For example, a node can receive many kinds of active messages (packets). Active messages (AM) have an 8-bit type field, to distinguish between protocols. A flooding protocol uses one AM type, while an ad hoc routing protocol uses another. ActiveMessageC, the component that signals the arrival of a packet, should not need to know what processing a protocol performs or whether an application supports a protocol. ActiveMessageC just delivers packets, and higher-level communication services respond to those they care about.

The traditional approach to this problem is to use function pointers or objects, which are dynamically registered as callbacks. In many cases, even though registered at run-time, the set of operations is known at compile-time. Thus these callbacks can be replaced by a dispatch table compiled into the executable, with two benefits. First, this allows better cross-function analysis and optimization, and secondly it conserves RAM, as no pointers or callback structures need to be stored.

Such a dispatch table could be built for the active message example by using a `switch` statement in ActiveMessageC. But this is very inflexible: any change to the protocols used in an application requires a change in a system component.

A better approach in TinyOS is to use the Dispatcher pattern. A Dispatcher invokes operations using a parameterized interface, based on a data identifier. In the case of ActiveMessageC, the interface is Receive and the identifier is the active message type field. ActiveMessageC is independent of what messages the application handles, or what processing those handlers perform. Adding a new handler requires a single wiring to ActiveMessageC. If an application does not wire a receive handler for a certain type, ActiveMessageC defaults to a null operation.

Another example of a Dispatcher is the scheduler of the Maté virtual machine that was implemented for TinyOS 1.x. Each instruction is a separate component that provides the MateBytecode interface. The scheduler executes a particular bytecode by dispatching to the instruction component using a parameterized MateBytecode interface. The instruction set can be easily changed by altering the wiring of the scheduler.

Applicable when

- A component needs to support an externally customisable set of operations.
- A primitive integer type can identify which operation to perform.
- The operations can all be implemented in terms of a single interface.

Structure

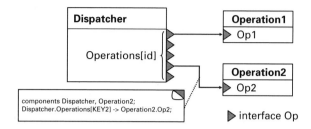

Participants

- **Dispatcher**: invokes its parameterized interface based on an integer type.
- **Operation**: implements the desired functionality and wires it to the dispatcher.

Sample code

CC1000ActiveMessageP is the component in the TI CC1000 radio stack responsible for dispatching received messages. The lower-level components in this radio stack pass all received messages (even those destined for other motes) to CC1000ActiveMessageP's SubReceive interface. This module dispatches messages intended for this mote to its parameterized Receive interface, and those intended for other motes to its parameterized Snoop interface:

```
module CC1000ActiveMessageP {
  provides interface Receive[am_id_t id];
  provides interface Receive as Snoop[am_id_t id];
  provides interface AMPacket;
  uses interface Receive as SubReceive;
}
implementation {
  event message_t* SubReceive.receive(message_t* msg, void* payload, uint8_t len
) {
    if (call AMPacket.isForMe(msg))
      return signal Receive.receive[call AMPacket.type(msg)](msg, payload, len);
    else
      return signal Snoop.receive[call AMPacket.type(msg)](msg, payload, len);
  }
  ...
}
```

Dispatchers can be wired too directly. This application receives two kinds of messages by wiring directly to ActiveMessageC:

```
configuration AppC {}
implementation {
  components AppP, ActiveMessageC;
  AppP.ClearIdMsg -> ActiveMessageC.Receive[AM_CLEARIDMSG];
  AppP.SetIdMsg -> ActiveMessageC.Receive[AM_SETIDMSG];
}
```

However, dispatchers are often encapsulated in a generic configuration to simplify their use. Rather than have a component wire to a parameterized interface, it wires to a generic configuration that takes the parameter as an argument. For example, rather than wire directly to ActiveMessageC to receive packets, applications instantiates an AMReceiverC:

```
generic configuration AMReceiverC(am_id_t amId) {
  provides interface Receive;
  provides interface Packet;
  provides interface AMPacket;
}
implementation {
  components ActiveMessageImplP as Impl;

  Receive = Impl.Receive[amId];
  Packet = Impl;
  AMPacket = Impl;
}
```

Listing 10.1 AMReceiverC

Rewriting AppC using AMReceiverC gives:

```
configuration AppC {}
implementation {
  components AppP,
    new AMReceiverC(AM_CLEARIDMSG) as ClearId,
    new AMReceiverC(AM_SETIDMSG) as SetId;

  AppP.ClearIdMsg -> ClearId;
  AppP.SetIdMsg -> SetId;
}
```

Known uses

The Active Messages networking layer (ActiveMessageC), and tree collection protocol (CollectionC) use a dispatcher for packet reception. They also provide a parameterized packet sending interface, so services can easily match packet sends to reception handlers.

The Atmel AT45DB family flash chip implementation uses a parameterized interface to map storage volume identifiers to storage volume characteristics (see the At45dbStorageManagerC component in Section 12.1.1).

The TinyOS 1.x Maté virtual machine uses a dispatcher to allow easy customization of instruction sets.

Consequences

By leaving operation selection to nesC wirings, the dispatcher's implementation remains independent of what an application supports. However, finding the full set of supported operations can require looking at many files. Sloppy operation identifier management can lead to dispatch problems. If two operations are wired with the same identifier, then a dispatch will call both, which may lead to resource conflicts, data corruption, or memory leaks from lost pointers. For example, the Receive interface uses a buffer swap mechanism to pass buffers between the radio stack and network services, in which the higher component passes a new buffer in the return value of the event. If two services are wired to a given Receive instance, only one of their pointers will be passed and the second will be lost. Wiring in this fashion is a compile-time warning in nesC, but it is still a common bug for novice TinyOS developers.

The key aspects of the dispatcher pattern are:

- It allows you to easily extend or modify the functionality an application supports: adding an operation requires a single wiring.
- It allows the elements of functionality to be independently implemented and re-used. Because each operation is implemented in a component, it can be easily included in many applications. Keeping implementations separate can also simplify testing, as the components will be smaller, simpler, and easier to pinpoint faults in. The nesC compiler will automatically inline small operations, or you can explicitly request inlining; thus this decomposition has no performance cost.
- It requires the individual operations to follow a uniform interface. The dispatcher is usually not well suited to operations that have a wide range of semantics. As all implementations have to meet the same interface, broad semantics leads to the interface being overly general, pushing error checks from compile-time to run-time. An implementor forgetting a run-time parameter check can cause a hard-to-diagnose system failure.

The compile-time binding of the operation simplifies program analysis and puts dispatch tables in the compiled code, saving RAM. Dispatching provides a simple way to develop programs that execute in reaction to their environment.

Related patterns

- Service Instance: a service instance creates many instances of an implementation of an interface, while a dispatcher selects between different implementations of an interface.
- Placeholder: a placeholder allows an application to select an implementation at compile-time, while a dispatcher allows it to select an implementation at run-time.
- Keyspace: the identifiers used to identify a Dispatcher's operation typically form a Global Keyspace.

10.2 Structural: Service Instance

Intent
Allows multiple users to have separate instances of a particular service, where the instances can collaborate efficiently. The basic mechanism for virtualizing services.

10.2 Structural: Service Instance

Motivation

Sometimes many components or subsystems need to use a system abstraction, but each user wants a separate instance of that service. We don't know how many users there will be until we build a complete application. Each instance requires maintaining some state, and the service implementation needs to access all of this state to make decisions.

For example, a wide range of TinyOS components need timers, for everything from network timeouts to sensor sampling. Each timer appears independent, but they all operate on top of a single hardware clock. An efficient implementation thus requires knowing the state of all of the timers. If the implementation can easily determine which timer has to fire next, then it can schedule the underlying clock resource to fire as few interrupts as possible to meet this lowest timer's requirement. Firing fewer interrupts allows the CPU to sleep more, saving energy and increasing lifetime.

The traditional object-oriented approach to this problem is to instantiate an object representing the service and use another class to coordinate state. The closest nesC equivalent would involve instantiating a generic module for each timer, and using another module to coordinate state. However, instantiating one module per timer leads to duplicated code and requires inter-module coordination in order to figure out how to set the underlying hardware clock. Inter-module coordination could be avoided by setting the clock to a fixed rate and maintaining a counter for each Timer, but this would be inefficient: timer fidelity requires firing at a high rate, but it wastes energy to fire at 1 KHz if the next timer is in four seconds.

The Service Instance pattern provides a solution to these problems. Using this pattern, each user of a service can have its own (virtual) instance, but instances share code and can access each other's state. A component following the Service Instance pattern provides its service via a parameterized interface; each user wires to a unique instance of the interface using unique. The underlying component receives the unique identity of each client in each command, and can use it to index into a state array. The component can determine at compile-time how many instances exist using the uniqueCount function and dimension the state array accordingly.

In most cases, components following the service instance pattern are made available to the user via a generic component that provides a non-parameterized interface, automating the use of unique. For instance, HilTimerMilliC follows the Service Instance pattern, but most TinyOS code uses the TimerMilliC generic component.

Applicable when

- A component needs to provide multiple instances of a service, but does not know how many until compile-time.
- Each service instance appears to its user to be independent of the others.
- The service implementation needs to be able to easily access the state of every instance.

Structure

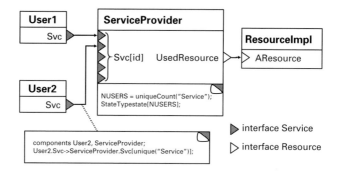

Participants

- **ServiceProvider**: allocates state for each instance of the service and coordinates underlying resources based on all of the instances.
- **ResourceImpl**: an underlying system resource that ServiceProvider multiplexes/demultiplexes service instances on.

Sample code

HilTimerMilliC uses a VirtualizeTimerC to present a Service Instance of millisecond precision timers. VirtualizeTimerC takes a single underlying timer and virtualizes it to *n* timers for others to use:

```
generic module VirtualizeTimerC( typedef precision_tag, int max_timers ) {
  provides interface Timer<precision_tag> as Timer[ uint8_t num ];
  uses interface Timer<precision_tag> as TimerFrom;
}
```

Listing 10.2 VirtualizeTimerC

It takes two parameters, a type for the Timer precision tag (TMilli, etc), and the number of timers (the number of service instances). Instantiating TimerMilliC to create a new timer thus involves many files:

1. TimerMilliC, which wires to HilTimerMilliC's parameterized interface with a call to unique passing UQ_TIMER_MILLI.
2. Timer.h, a header file that defines the unique key as UQ_TIMER_MILLI.
3. HilTimerMilliC, which instantiates a VirtualizeTimerC with a call to uniqueCount passing UQ_TIMER_MILLI.
4. VirtualizeTimerC, which provides *n* virtualized timers.
5. The underlying per-platform components that provide the single Timer to be virtualized.

10.2 Structural: Service Instance

Known uses

Most TinyOS services follow the Service Instance pattern, though most services (e.g. TimerMilliC, BlockStorageC from Chapter 6 or Atm128I2CMasterC from Chapter 13) hide it behind a programmer-friendly generic component.

The generic arbiter implementations such as FcfsArbiterC (Chapter 11) expose the Service Instance pattern to the programmer, as does VirtualizeTimerC.

In a similar vein, the epidemic dissemination protocol Drip (available via the DisseminatorC generic component) uses the service instance pattern to maintain epidemic state for each disseminated value.

Consequences

The key aspects of the Service Instance pattern are:

- It allows many components to request independent instances of a common system service: adding an instance requires a single wiring.
- It controls state allocation, so the amount of RAM used is scaled to exactly the number of instances needed, conserving memory while preventing run-time failures due to many requests exhausting resources.
- It allows a single component to coordinate all of the instances, which enables efficient resource management and coordination.

Because the pattern scales to a variable number of instances, the cost of its operations may scale linearly with the number of users. For example, if setting the underlying clock interrupt rate depends on the timer with the shortest remaining duration, an implementation might determine this by scanning all of the timers, an $O(n)$ operation.

If many users require an instance of a service, but each of those instances are used rarely, then allocating state for each one can be wasteful. The other option is to allocate a smaller amount of state and dynamically allocate it to users as needs be. This can conserve RAM, but requires more RAM per real instance (client IDs need to be maintained), imposes a CPU overhead (allocation and deallocation), can fail at run-time (if there are too many simultaneous users), and assumes a reclamation strategy (misuse of which would lead to leaks). This long list of challenges makes the Service Instance an attractive – and more and more commonly used – way to efficiently support application requirements. There are situations, however, when a component internally re-uses a single service instance for several purposes: for example, the Maté virtual machine code propagation component uses a single timer instance for several different timers which never operate concurrently.

Related patterns

- Dispatcher: a service instance creates many instances of an implementation of an interface, while a dispatcher selects between different implementations of an interface.
- Keyspace: a Service Instance's instance identifiers form a Local Keyspace.

10.3 Namespace: Keyspace

Intent

Provide namespaces for referring to protocols, structures, or other entities in a program.

Motivation

A typical sensor network program needs namespaces for the various entities it manages, such as protocols, data types, or structure instances. Limited resources mean names are usually stored as small integer keys.

For data types representing internal program structures, each instance must have a unique name, but as they are only relevant to a single mote, the names can be chosen freely. These *local* namespaces are usually dense, for efficiency. The Service Instance pattern (Section 10.2) uses a local namespace to identify instances. In contrast, communication requires a shared, *global* namespace: two motes/applications must agree on an element's name. As a mote may only use a few elements, global namespaces are typically sparse. The Dispatcher pattern (Section 10.1) uses a global namespace to select operations.

The Keyspace patterns provide solutions to these problems. Using these patterns, programs can refer to elements using identifiers optimized for their particular use. Components using the Keyspace patterns often take advantage of a parameterized interface, in which the parameter is an element in a Keyspace. Local Keyspaces are designed for referring to local data structures (e.g. arrays) and are generated with unique; Global Keyspaces are designed for communication and use global constants (normally defined using **enum**).

The AM identifiers used to distinguish message types form a Global Keyspace. ActiveMessageC uses these AM identifiers in conjunction with a Dispatcher to execute the appropriate message handlers found in other components. Local Keyspaces have other uses than identifying clients in the Service Instance pattern: the Maté virtual machine uses a Local Keyspace to identify locks corresponding to resources used by Maté programs. These lock identifiers are allocated with unique as the Maté virtual machine can be compiled with varying sets of resources.

Applicable when

- A program must keep track of a set of elements or data types.
- The set is known and fixed at compile-time.

Sample code

The AM identifiers for the SoundLocalizer application (Chapter 13) are defined as global constants:

```
enum {
  AM_COORDINATION_MSG = 101,
  AM_DETECTION_MSG = 102
};
```

10.3 Namespace: Keyspace

which are then used when instantiating the AMSenderC and AMReceiverC generic components that allow SoundLocalizer to send and receive messages:

```
components SynchronizerC;
components new AMReceiverC(AM_COORDINATION_MSG) as CReceive;
components new AMReceiverC(AM_DETECTION_MSG) as DReceive;
components new AMSenderC(AM_DETECTION_MSG) as DSend;

SynchronizerC.RCoordination -> CReceive;
SynchronizerC.RDetection -> DReceive;
SynchronizerC.SDetection -> DSend;
```

Within the radio stack, these AM identifiers are placed into the type field of outgoing messages. This same type field is used to dispatch received messages:

```
command error_t AMSend.send[am_id_t id](am_addr_t addr,
                                        message_t *amsg,
                                        uint8_t len) {
  radio_header_t* header = getHeader(amsg);
  header->type = id;
  ...
}

message_t *received(message_t *msg, void *payload, uint8_t len) {
  radio_header_t* header = getHeader(msg);
  msg = signal Receive.receive[header->type](msg, payload, len);
  ...
}
```

The TinyOS 1.x Maté lock subsystem identifies locks by small integers:

```
module MLocks {
  provides interface MateLocks as Locks;
}
implementation {
  MateLock locks[MATE_LOCK_COUNT];

  command void Locks.lock(MateContext* uint8_t lockNum) {
    locks[lockNum].holder = context;
    context->heldSet[lockNum / 8] |= 1 << (lockNum % 8);
  }
  ...
```

Locks are allocated in components providing shared resources:

```
module OPgetsetvar1M { ... } // a shared variable
implementation {
  typedef enum {
    MATE_LOCK_1_0 = unique("MateLock"),
    MATE_LOCK_1_1 = unique("MateLock"),
```

```
    } LockNames;
  ...
  module OPbpush1M { ... } // a shared buffer
  implementation {
    typedef enum {
      MATE_BUF_LOCK_1_0 = unique("MateLock"),
      MATE_BUF_LOCK_1_1 = unique("MateLock"),
    } BufLockNames;
    ...
```

and uniqueCount is used to find the total number of locks:

```
enum {
  MATE_LOCK_COUNT = uniqueCount("MateLock")
};
```

Known uses

Many components use Local Keyspaces: they are a fundamental part of the Service Instance pattern. See for example the timer service, HilTimerMilliC, or the client identifiers of many lower-level system components such as Atm128AdcC (Atmel ATmega128 A/D converter) or Stm25pSectorC (ST M25P flash chip).

Maté uses a Local Keyspace to keep track of Maté shared resource locks (see above).

Active Messages (ActiveMessageC) uses a Global Keyspace for Active Message types.

The TinyOS storage abstractions use a Global Keyspace to identify flash-chip volumes. In this case, the global constants are picked by an external tool (tos-storage-at45db for the Atmel AT45DB chip family) rather than explicitly by the programmer. This relieves the programmer of the burden of picking values, while still allowing several applications to use the same volumes.

The TinyDB sensor-network-as-database application [20] uses a Global Keyspace for its attributes; in this case, however, the keyspace is composed of strings, which are then mapped to a Local Keyspace using a table.

Consequences

Keyspaces allow a component to refer to data items or types through a parameterized interface. In a Local Keyspace, unique ensures that every element has a unique identifier. Global Keyspaces can also have unique identifiers, but this requires external namespace management.

As Local Keyspaces are generated with unique, mapping names to keys (e.g. for debugging purposes) is not obvious. The nesC constant generator, `ncg`, can be used to extract this information.

Keyspaces are rarely used in isolation; they support other patterns such as Dispatcher and Service Instance.

Related patterns

- Keymap: two Keyspaces are often related, e.g. one Service Instance may be built on top of another, requiring a mapping between two Keyspaces. The Keymap pattern provides an efficient way of implementing such maps.

- Service Instance: the identifiers used to identify individual services form a Local Keyspace.
- Dispatcher: the identifiers used by a dispatcher are typically taken from a Global Keyspace.

10.4 Namespace: Keymap

Intent

Map keys from one keyspace to another. Allows you to translate global, shared names to local, optimized names, or to efficiently subset another keyspace.

Motivation

Mapping between namespaces is often useful: it allows motes to use a global, sparse namespace for easy cross-application communication and an internal, compact namespace for efficiency.

The Drip epidemic dissemination protocol uses the Keyspace and Keymap patterns to allow a user to configure parameters at run-time. A component registers a parameter with the DisseminatorC generic component with a Global Keyspace, so it can be named in an application-independent manner. The user modifies a parameter by sending a key-value pair using an epidemic protocol, which distributes the change to every mote. Drip maintains a special "trickle" timer for each key-value pair, accessed via a Local Keyspace for efficiency. A Keymap maps the global key to the local key.

Keymaps are also useful for mapping between two local keyspaces, when some service, based on the Service Instance pattern, accesses a subset of the resources provided by another service, also based on the Service Instance pattern.

For instance, the Atmel AT45db implementation of the BlockStorageC, LogStorageC, and ConfigStorageC components (Section 6.5) identify their respective clients with their own local keyspace. All three are built upon the lower-level At45dbC component, which also identifies its clients with its own local keyspace. Each client of, e.g. BlockStorageC is thus indirectly a client of At45dbC. A keymap maps BlockStorageC's client identifiers to At45dbC's client identifiers.

Maps could be implemented using a table and some lookup code. However, this has several problems. If we want to store this table in ROM, then it must be initialized in one place. However, this conflicts with the desire to specify keys in separate components (either with unique or with constants). If the table is stored in RAM, then keys can be specified in separate components, but RAM is in very short supply on motes. Finally, keys of Global Keyspaces are sparse, so the resulting tables would be large and waste space.

Instead, we can use nesC's wiring to build Keymaps. By mapping a parameterized interface indexed with one key to another parameterized interface indexed by a second key, we can have the nesC compiler generate the map at compile-time. Additionally, as the map exists as an automatically generated switch statement, it uses no RAM.

Design patterns

Applicable when

- An application uses global identifiers for communication (or other purposes) and wishes to map them to local identifiers for efficiency.
- Two services are implemented following the Service Instance pattern, and the first service needs an instance of the second service for each of its own instances.
- The identifiers are integer constants.
- The map is known at compile-time.

Structure

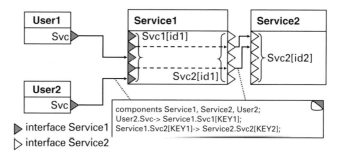

Participants

- **Service1**: service accessed via key 1, dependent on Service2.
- **Service2**: service accessed via key 2.

Sample code

The DisseminatorC generic component creates a new key-value pair for use with the Drip dissemination protocol:

```
enum { DISS_CONFIG = 42 };
components new DisseminatorC(struct configuration, DISS_CONFIG);
```

DisseminatorC is implemented using the underlying DisseminationEngineP component, which identifies key-value pairs by the global key value, and the DisseminationTimerP which identifies "trickle" timer components using a local keyspace. DisseminatorC builds the keymap that connects the global drip keys to the local "trickle" timer keys:

```
generic configuration DisseminatorC(typedef t, uint16_t key) {
    provides interface DisseminationValue<t>;
    provides interface DisseminationUpdate<t>;
}
implementation {
    enum { TIMER_ID = unique(UQ_DISSEMINATION_TRICKLETIMER) };

    components DisseminationEngineP;
    components DisseminationTimerP;
```

```
    DisseminationEngineP.TrickleTimer[key] ->
      DisseminationTimerP.TrickleTimer[TIMER_ID];
    ...
  }
```

In this example, a user can generate a new configuration value, and distribute it based on the `DISS_CONFIG` key. DisseminatorEngineP uses the global key to refer to the value, but DisseminationTimerP can use a local key to refer to the state it maintains for its "trickle" timers. The wiring compiles down to a simple switch statement that calls DisseminationTimerP with the proper local key.

BlockStorageC provides a simple read/write abstraction over a flash chip volume. Internally, BlockStorageC is implemented using a Service Instance pattern provided by the BlockStorageP component. Each client of BlockStorageP is also a client of At45dbC (this simplifies BlockStorageC's implementation). As At45dbC has other clients, BlockStorageC must map BlockStorageP identifiers to At45dbC identifiers:

```
generic configuration BlockStorageC(volume_id_t volid) {
  provides interface BlockWrite;
  provides interface BlockRead;
}
implementation {
  enum {
    BLOCK_ID = unique(UQ_BLOCK_STORAGE),
    AT45DB_ID = unique(UQ_AT45DB)
  };

  components BlockStorageP, At45dbC;
  BlockWrite = BlockStorageP.BlockWrite[BLOCK_ID];
  BlockRead = BlockStorageP.BlockRead[BLOCK_ID];
  BlockStorageP.Resource[BLOCK_ID] -> At45dbC.Resource[AT45DB_ID];
  ...
}
```

Let's assume that when BLOCK_ID is two, AT45DB_ID is four. Then, when a request comes in for `BlockRead[2]` (or `BlockWrite[2]`) interface, BlockStorageP makes a request on its `Resource[2]` interface. The keymap translates this request into a request on `At45dbC.Resource[4]`, as desired.

Known uses

The Drip dissemination protocol – described above – uses a Keymap.

The TinyOS storage system for the ST M25P and Atmel AT45DB family flash chips (the latter described above) use a Keymap.

Consequences

A Keymap uses nesC wiring to allow components to transparently map between different keyspaces. As with Keyspaces, the Keymap must be fixed at compile-time.

A Keymap translates into a `switch` at compile-time. It thus doesn't use any RAM; its speed depends on the behaviour of the C compiler used to compile nesC's output.

Keymaps only support mapping between integers. If you need, e.g. to map from strings to a Local Keyspace, you will need to build your own map.

Related patterns

- Keyspace: A Keymap establishes a map from one keyspace to another.

10.5 Structural: Placeholder

Intent
Easily change which implementation of a service an entire application uses. Prevent inadvertent inclusion of multiple, incompatible implementations.

Motivation
Many TinyOS systems and abstractions have several implementations. For example, there are many ad hoc tree routing protocols, including two (Ctp, Lqi) in the TinyOS core, but they all expose the same interfaces (StdControl, Send, Receive, etc.). The standardized interface allows applications to use any of the implementations without code changes. Simpler abstractions can also have multiple implementations. For example, the LedsC component actually turns the LEDs on and off, while the NoLedsC component, which provides the same interface, has null operations. During testing, LedsC is useful for debugging, but in deployment it is a significant energy cost and usually replaced with NoLedsC.

Sometimes, the decision of which implementation to use needs to be uniform across an application. For example, if a hypothetical network health monitoring subsystem wires to Ctp, while an application uses Lqi, two routing trees will be built, wasting resources. As every configuration that wires to a service names it, changing the choice of implementation in a large application could require changing many files. Some of these files, such as the network health monitor might be part of the system; an application writer should not have to modify them.

One option is for every implementation to use the same component name, and put them in separate directories. Manipulating the nesC search order allows an application to select which version to use. However, this forces every implementation of the placeholder into a separate directory and precludes the possibility of including two implementations, even if they can interoperate.

The Placeholder pattern offers a solution. A placeholder configuration represents the desired service through a level of naming indirection. All components that need to use the service wire to the placeholder. The placeholder itself is just "a pass through" of the service's interfaces to a particular implementation. These implementations should have a name that is related to the placeholder, e.g. RandomLfsrC and RandomMlcgC are two implementations of RandomC. Placeholders for system components are provided by TinyOS itself, but can be overridden by creating a replacement placeholder. Components can still wire to a specific implementation by name. As the level of indirection is solely in terms of names – there is no additional code generated – it imposes no CPU overhead.

10.5 Structural: Placeholder

Applicable when

- A component or service has multiple, possibly mutually exclusive, implementations.
- Many subsystems and parts of your application need to use this component/service.
- You need to easily switch between the implementations.

Structure

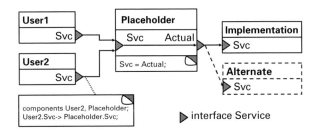

Participants

- **Placeholder**: the component that all other components wire to. It encapsulates the implementation and exports its interfaces with pass-through wiring. It has the same signature as the Implementation component.
- **Implementation**: the specific version of the component.

Sample code

The Telos platform uses a Placeholder to map ActiveMessageC to its radio stack, CC2420ActiveMessageC:

```
configuration ActiveMessageC {
  provides {
    interface Init;
    interface SplitControl;

    interface AMSend[uint8_t id];
    interface Receive[uint8_t id];
    interface Receive as Snoop[uint8_t id];

    interface Packet;
    interface AMPacket;
    interface PacketAcknowledgements;
  }
}
implementation {
  components CC2420ActiveMessageC as AM;

  Init          = AM;
  SplitControl  = AM;

  AMSend        = AM;
  Receive       = AM.Receive;
  Snoop         = AM.Snoop;
```

```
  Packet                = AM;
  AMPacket              = AM;
  PacketAcknowledgements = AM;
}
```

Listing 10.3 Telos ActiveMessageC

The Ctp and Lqi tree-routing protocols provide CtpCollectionC and LqiCollectionC components in separate directories. They also provide a placeholder CollectionC component that wires to CtpCollectionC or LqiCollectionC. Components that wish to use the default tree-collection wire to CollectionC, leaving the choice of tree-routing protocol to the application. Components that wish to use a specific tree-collection algorithm wire, e.g. to CtpCollectionC.

Known uses

Hardware-independent abstractions which have multiple implementations are often placeholders. For instance, each platform must map the ActiveMessageC and HilTimerMilliC components to a platform-specific implementation. In many cases, like CC2420ActiveMessageC, this implementation is specific to a particular radio chip but portable across mote platforms. Leaving the mapping of ActiveMessageC to the platform can be important if a mote has two radios and one should be used as the default.

CollectionC is the placeholder for tree-collection routing protocols, DisseminatorC is the placeholder for key-value dissemination protocols, and RandomC is the placeholder for random-number generation.

Consequences

The key aspects of the Placeholder pattern are:

- Establishes a global name that users of a common service can wire to.
- Allows you to specify the implementation of the service on an application-wide basis.
- Does not require every component to use the Placeholder's implementation.

By adding a level of naming indirection, a Placeholder provides a single point at which you can choose an implementation. Placeholders create a global namespace for implementation-independent users of common system services. As using the Placeholder pattern generally requires every component to wire to the Placeholder instead of a concrete instance, incorporating a Placeholder into an existing application can require modifying many components. However, the nesC compiler optimises away the added level of wiring indirection, so a Placeholder imposes no run-time overhead. The Placeholder supports flexible composition and simplifies use of alternative service implementations.

Related patterns

- Dispatcher: a placeholder allows an application to select an implementation at compile-time, while a dispatcher allows it to select an implementation at run-time.

- Facade: a placeholder allows easy selection of the implementation of a group of interfaces, while a facade allows easy use of a group of interfaces. An application may well connect a placeholder to a facade.

10.6 Structural: Facade

Intent

Provides a unified access point to a set of inter-related services and interfaces. Simplifies use, inclusion, and composition of the subservices.

Motivation

Complex system components, such as a filesystem or networking abstraction, are often implemented across many components. Higher-level operations may be based on lower-level ones, and a user needs access to both. Complex functionality may be spread across several components. Although implemented separately, these pieces of functionality are part of a cohesive whole that we want to present as a logical unit.

For example, the Matchbox filing system from TinyOS 1.x provides interfaces for reading and writing files, as well as for metadata operations such as deleting and renaming. Separate modules implement each of the interfaces, depending on common underlying services such as reading blocks.

One option would be to put all of the operations in a single, shared interface. This raises two problems. First, the nesC wiring rules mean that a component that wants to use *any* command in the interface has to handle *all* of its events. In the case of a file system, all the operations are split-phase; having to handle a half dozen events (readDone, writeDone, openDone, etc.) merely to be able to delete a file is hardly usable. Second, the implementation cannot be easily decomposed into separate components without introducing internal interfaces, as the top-level component will need to call out into the subcomponents. Implementing the entire subsystem as a single huge component is not easy to maintain.

Another option is to export each interface in a separate component (e.g. MatchboxRead, MatchboxWrite, MatchboxRename, etc.). This increases wiring complexity, making the abstraction more difficult to use. For a simple open, read, and write sequence, the application would have to wire to three different components. Additionally, each interface would need a separate configuration to wire it to the subsystems it depends on, increasing clutter in the component namespace. The implementer needs to be careful with these configurations, to prevent inadvertent double-wirings.

The Facade pattern provides a better solution to this problem. The Facade pattern provides a uniform access point to interfaces provided by many components. A Facade is a nesC configuration that defines a coherent abstraction boundary by exporting the interfaces of several underlying components. Additionally, the Facade can wire the underlying components, simplifying dependency resolution.

A nesC Facade has strong resemblances to the object oriented pattern of the same name. The distinction lies in nesC's static model. An object-oriented Facade instantiates its subcomponents at run-time, storing pointers and resolving operations through another level of call indirection. In contrast, as a nesC Facade is defined through naming (pass through wiring) at compile-time, there is no run-time cost.

Applicable when

- An abstraction, or series of related abstractions, is implemented across several separate components.
- It is preferable to present the abstraction in whole rather than in parts.

Structure

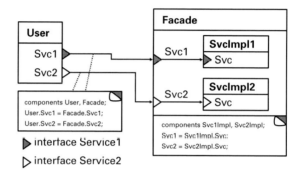

Participants

- **Facade**: the uniform presentation of a set of related services.
- **SvcImpl**: the separate implementations of each service composing the Facade.

Sample 1.x code

The Matchbox filing system uses a Facade to present a uniform filesystem abstraction. File operations are all implemented in different components, but the top-level Matchbox configuration provides them in a single place. Each of these components depends on a wide range of underlying abstractions, such as a block interface to non-volatile storage; Matchbox wires them appropriately, resolving all of the dependencies.

```
configuration Matchbox {
  provides {
    interface FileRead[uint8_t fd];
    interface FileWrite[uint8_t fd];
    interface FileDir;
    interface FileRename;
    interface FileDelete;
  }
}
implementation {
  // File operation implementations
  components Read, Write, Dir, Rename, Delete;

  FileRead = Read.FileRead;
  FileWrite = Write.FileWrite;
```

```
        FileDir = Dir.FileDir;
        FileRename = Rename.FileRename;
        FileDelete = Delete.FileDelete;
        // Wiring of operations to sub-services omitted
    }
```

Listing 10.4 The Matchbox facade

Sample 2.x code

The CC2420 radio stack is broken up into three call paths: control, transmission, and reception. The top-level CC2420CsmaC component presents these three paths together as a single abstraction using a Facade:

```
configuration CC2420CsmaC {
    provides interface Init;
    provides interface SplitControl;

    provides interface Send;
    provides interface Receive;
    provides interface PacketAcknowledgements as Acks;

    uses interface AMPacket;
}
implementation {
    components CC2420CsmaP as CsmaP;
    Init = CsmaP;
    SplitControl = CsmaP;
    Send = CsmaP;
    Acks = CsmaP;
    AMPacket = CsmaP;

    components CC2420ControlC;
    Init = CC2420ControlC;
    AMPacket = CC2420ControlC;
    CsmaP.Resource -> CC2420ControlC;
    CsmaP.CC2420Config -> CC2420ControlC;

    components CC2420TransmitC;
    Init = CC2420TransmitC;
    CsmaP.SubControl -> CC2420TransmitC;
    CsmaP.CC2420Transmit -> CC2420TransmitC;
    CsmaP.CsmaBackoff -> CC2420TransmitC;

    components CC2420ReceiveC;
    Init = CC2420ReceiveC;
    Receive = CC2420ReceiveC;
    CsmaP.SubControl -> CC2420ReceiveC;

    components RandomC;
    CsmaP.Random -> RandomC;
```

```
    components LedsC as Leds;
    CsmaP.Leds -> Leds;
}
```

Listing 10.5 The CC2420CsmaC uses a Facade

Known uses
Stable, commonly used abstract boundaries such as radio stacks (CC2420CsmaC, CC1000CsmaRadioC) and storage (BlockStorageC) often use a Facade. This allows them to implement complex abstractions in smaller, distinct parts, which simplifies code.

Consequences
The key aspects of the Facade pattern are:

- Provides an abstraction boundary as a set of interfaces. A user can easily see the set of operations the abstraction support, and only needs to include a single component to use the whole service.
- Presents the interfaces separately. A user can wire to only the needed parts of the abstraction, but be certain everything underneath is composed correctly.

A Facade is not always without cost. Because the Facade names all of its subparts, they will all be included in the application. While the nesC compiler attempts to remove unreachable code, this analysis is necessarily conservative and may end up keeping much useless code. In particular, unused interrupt handlers are never removed, so all the code reachable from them will be included every time the Facade is used. If you expect applications to only use a very narrow part of an abstraction, then a Facade can be wasteful.

Related patterns

- Placeholder: a placeholder allows easy selection of the implementation of a group of interfaces, while a Facade allows easy use of a group of interfaces. An application may well connect a placeholder to a Facade.

10.7 Behavioral: Decorator

Intent
Enhance or modify a component's capabilities without modifying its implementation. Be able to apply these changes to any component that provides the interface.

Motivation
We often need to add extra functionality to an existing component, or to modify the way it works without changing its interfaces. For instance, the standard LogStorageC component provides a LogWrite interface to log data to a region of flash memory. In some circumstances, we would like to introduce a RAM write buffer on top of the interface. This would reduce the number of times the application talks to the flash chip, improving performance and conserving energy.

Adding a buffer to the LogStorageC component forces all logging applications to allocate the buffer. As some applications may not able to spare the RAM, this is undesirable. Providing two versions, buffered and unbuffered, replicates code, reducing reuse, and increasing the possibility of incomplete bug fixes. It is possible that several implementers of the interface – any component that provides LogWrite – may benefit from the added functionality. Having multiple copies of the buffering version, spread across several services, further replicates code.

There are two traditional object-oriented approaches to this problem: inheritance, which defines the relationship at compile-time through a class hierarchy, and decorators, which define the relationship at run-time through encapsulation. As nesC is not an object-oriented language, and has no notion of inheritance, the former option is not possible. Similarly, run-time encapsulation is not readily supported by nesC's static component composition model and imposes overhead in terms of pointers and call forwarding. However, we can use nesC's component composition and wiring to provide a compile-time version of the Decorator.

A Decorator component is typically a generic module that provides and uses the same interface type, such as LogWrite. The provided interface adds functionality on top of the used interface. For example, we show a BufferedLogC component that sits on top of a LogWrite provider. It implements its additional functionality by aggregating several writes to BufferedLogC writes into a single write to the underlying LogWrite interface.

Using a Decorator can have further benefits. In addition to augmenting existing interfaces, they can introduce new ones that provide alternative abstractions. For example, BufferedLogC provides a synchronous (not split phase) FastLog interface; a call to FastLog writes directly into the buffer.

Finally, separating added functionality into a Decorator allows it to apply to any implementation. For example, a packet send queue Decorator can be interposed on top of any networking abstraction that provides the Send interface; this allows flexible interpositioning of queues and queueing policies in a networking system.

Applicable when

- You wish to extend the functionality of an existing component without changing its implementation, or
- You wish to provide several variants of a component without having to implement each possible combination separately.

Structure

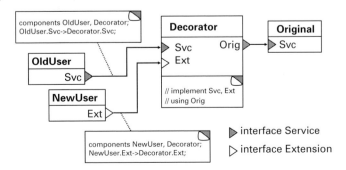

Participants

- **Original**: the original service.
- **Decorator**: the extra functionality added to the service.

Sample code

The standard LogWrite interface (Section 6.5) includes split-phase erase, append and sync operations. BufferedLogC adds buffering to the LogData operations, and, additionally, supports a FastLog interface with a non-split-phase append operation (for small writes only):

```
generic module BufferedLogC(unsigned int bufsize) {
  provides interface LogWrite;
  provides interface FastLog;
  uses interface LogWrite as UnbufferedLog;
}
implementation {
  uint8_t buffer1[bufsize], buffer2[bufsize];
  uint8_t *buffer;
  command result_t FastLog.append(void *data, storage_len_t n) {
    if (bufferFull()) {
      call UnbufferedLog.append(buffer, offset);
      // ... switch to other buffer ...
    }
    // ... append to buffer ...
  }
```

Known uses

The TransformAlarmC and TransformCounterC generic components transform the precision and width of Alarm and Counter interfaces.

The ArbitratedReadC generic component adds automatic resource arbitration to a Read interface, and ArbitratedReadStreamC does the same for a ReadStream interface.

Consequences

Applying a Decorator allows you to extend or modify a component's behavior though a separate component: the original implementation can remain unchanged. Additionally, the Decorator can be applied to any component that provides the interface. To allow reuse, Decorators are normally generic components.

In most cases, a decorated component should not be used directly, as the Decorator is already handling its events. The Placeholder pattern (Section 10.5) can be used to help ensure this.

Additional interfaces are likely to use the underlying component, creating dependencies between the original and extra interfaces of a Decorator. For instance, in BufferedLogC, FastLog uses UnbufferedLog, so concurrent requests to FastLog and Log are likely to conflict: only one can access the UnbufferedLog at once.

Decorators are a lightweight but flexible way to extend component functionality. Interpositioning is a common technique in building networking stacks, and Decorators enable this style of composition.

Related patterns

- Adapter: An Adapter presents the existing functionality of a component with a different interface, rather than adding additional functionality and preserving the current interface.

10.8 Behavioral: Adapter

Intent

Convert the interface of a component into another interface, without modifying the original implementation. Allow two components with different interfaces to interoperate.

Motivation

Sometimes, a piece of functionality offered by a component with one interface needs to be accessed by another component via a different interface. For instance, the low-level Alarm interface is efficient but signals its "fired" event in an interrupt handler, exposing its users to the complexity of interrupt-driven programming. Using the Timer interface is preferable in most cases, as it is much easier to write correct task-level code.

Manually implementing Timer instead of Alarm for every timer, or even once for every platform, is undesirable: some of the repeated code would likely contain bugs and lead to maintenance problems if the Timer or Alarm interfaces change. Instead, the AlarmToTimerC Adapter implements a Timer interface in terms of Alarm's operations.

An Adapter is a component (normally generic) which provides an interface of type A, e.g. Timer, and uses an interface of type B, e.g. Alarm, and implements the operations of A in terms of those of B. An Adapter may also need to implement functionality not provided by the B interface, e.g. Timer provides periodic and one-shot timing events while Alarm only has one-shot events. More generally, an Adapter may provide several interfaces $A_1, \ldots, A_n$ and implement them in terms of several used interfaces $B_1, \ldots, B_m$.

Applicable when

- You wish to provide the functionality of an existing component with a different interface.

Structure

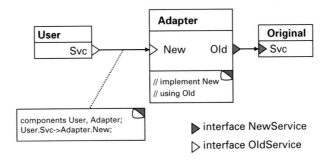

Participants

- **Original**: the original service.
- **Adapter**: implements the new interface in terms of the functionality offered by the old.

Sample code

The AlarmToTimerC converts the interrupt-driven (async) Alarm interface to a task-driven (not async) Timer interface, and implements Timer's periodic events in terms of Alarm's one-shot events:

```
generic module AlarmToTimerC(typedef precision_tag)
{
  provides interface Timer<precision_tag>;
  uses interface Alarm<precision_tag,uint32_t>;
}
implementation {
  uint32_t m_dt;
  bool m_oneshot;

  void start(uint32_t t0, uint32_t dt, bool oneshot) {
    m_dt = dt;
    m_oneshot = oneshot;
    call Alarm.startAt(t0, dt);
  }

  command void Timer.startPeriodicAt(uint32_t t0, uint32_t dt) {
    start(t0, dt, FALSE);
  }

  command void Timer.startOneShotAt(uint32_t t0, uint32_t dt) {
    start(t0, dt, TRUE);
  }

  task void fired() {
    if(m_oneshot == FALSE)
      start(call Alarm.getAlarm(), m_dt, FALSE);
    signal Timer.fired();
  }

  async event void Alarm.fired() {
    post fired();
  }
  ...
}
```

Listing 10.6 AlarmToTimerC implementation

Known uses

In TinyOS, hardware resources such as A/D converters are presented by a hardware abstraction layer (HAL) which offers high-level, but hardware-specific interfaces and a hardware independent layer (HIL) which offers high-level, platform-independent interfaces (see Chapter 12). The HIL layer is typically an Adapter over the HAL layer. For example, see the AdcP A/D converter component for the ATmega128.

TinyOS's provides a number of Adapters to simplify implementation of a mote's timer subsystem, including AlarmToTimerC (seen above), and CounterToLocalTimeC (convert a Counter to a LocalTime interface).

Consequences

An Adapter allows a component to be reused in circumstances other than initially planned for, without changing the original implementation.

In many cases, a component used with an Adapter cannot be used independently in the same application, as the Adapter will already be handling its events. As with the Decorator, the Placeholder pattern (Section 10.5) can help ensure this.

An Adapter can be used to adapt many different implementations of its used interfaces if it doesn't embody assumptions or behavior specific to a particular adapted component. Like Decorators, Adapters are normally generic components to allow reuse.

Adding an additional layer to convert between interfaces may increase the application's resource consumption (ROM, RAM, and execution time).

Related patterns

- Decorator: A Decorator adds functionality to an existing component while preserving its original interface. An Adapter presents existing (and possibly additional) functionality via a different interface.

11 Concurrency

So far, the code we've looked at has been split-phase and runs in tasks (synchronous). This programming model is sufficient for almost all application-level code. High-performance applications and low-level device drivers sometimes require additional functionality and concurrency models. This chapter describes two such additional mechanisms: asynchronous code and resource locks. Asynchronous code is a feature of the nesC language, while resource locks are a set of TinyOS components and mechanisms.

11.1 Asynchronous code

Tasks allow software components to emulate the split-phase behavior of hardware. But they have much greater utility than that. They also provide a mechanism to manage preemption in the system. Because tasks run atomically with respect to one another, code that runs only in tasks can be rather simple: there's no danger of another execution suddenly taking over and modifying data under you. However, interrupts do exactly that: they interrupt the current execution and start running preemptively.

11.1.1 The async keyword

As we saw in Chapter 5, nesC distinguishes between synchronous (sync) and asynchronous (async) code. Commands and events that can run preemptively from interrupt handlers (and therefore asynchronously with regards to tasks), must be labeled with the **async** keyword, both in the interface where the command or event is declared and in the module where the command or event is implemented. As a result, an async command or event, or a function reachable from an async command or event can only call or signal async commands and events (nesC will tell you when you break this rule). This rule means that it's clear from looking at an interface or module which code is synchronous and which is asynchronous. For example, the Send interface is purely synchronous as no commands or events are marked with **async**:

```
interface Send {
  command error_t send(message_t* msg, uint8_t len);
  event void sendDone(message_t* msg, error_t error);
```

```
    command error_t cancel(message_t* msg);
    command void* getPayload(message_t* msg);
    command uint8_t maxPayloadLength(message_t* msg);
}
```

Listing 11.1 The Send interface

In contrast, the Leds interface is purely asynchronous:

```
interface Leds {
  async command void led0On();
  async command void led0Off();
  async command void led0Toggle();

  ... more commands declared with async ...
}
```

Listing 11.2 The Leds interface

All interrupt handlers are automatically async, and so they cannot include any sync functions in their call graph. The one and only way that an interrupt handler can execute a sync function is to post a task. A task post is allowed from an async context, but the resulting task runs in a synchronous context.

For example, consider a packet layer on top of a UART. When the UART receives a byte, it signals an interrupt. In the interrupt handler, software reads the byte out of the data register and puts it in a buffer. When the last byte of a packet is received, the software needs to signal packet reception. But the receive event of the Receive interface is sync. So in the interrupt handler of the final byte, the component posts a task to signal packet reception.

11.1.2 The cost of async

This raises the question: If tasks introduce latency, why use them at all? Why not make everything async? The reason is simple: race conditions, in particular *data races*. The basic problem with preemptive execution is that it can modify state underneath an ongoing computation, which can cause a system to enter an inconsistent state. For example, consider this command, toggle, which flips the state bit and returns the old one:

```
bool state;
async command bool toggle() {
  if (state == 0) {
    state = 1;
    return 1;
  }
```

```
    if (state == 1) {
      state = 0;
      return 0;
    }
  }
```

Listing 11.3 Toggling a state variable

Now imagine this execution, which starts with state = 0:

```
toggle()
  state = 1;
    -> interrupt
    toggle()
      state = 0
      return 0;
  return 1;
```

Listing 11.4 A call sequence that could corrupt a variable

In this execution, when the first toggle returns, the calling component will think that state is equal to 1. But the last assignment (in the interrupt) was to 0.

This problem can be much worse when a single statement can be interrupted. For example, on micaz or Telos motes, writing or reading a 32-bit number takes more than one instruction. It's possible that an interrupt executes in between two instructions, so that part of the number read is of an old value while another part is of a new value.

This problem – data races – is particularly pronounced with state variables. For example, imagine this is a snippet of code from AMStandard, the basic packet abstraction in TinyOS 1.x, with a bunch of details omitted. The state variable indicates whether the component is busy.

```
command result_t SendMsg.send ... {
  if (!state) {
    state = TRUE;
    // send a packet
    return SUCCESS;
  }
  return FAIL;
}
```

Listing 11.5 State transition that is not async-safe

If this command were async, then it's possible between the conditional `if (!state)` and the assignment `state = TRUE` that another component jumps in and tries to send as well. This second call will see state to be false, set state to true, start a send and return SUCCESS. But then the first caller will result, send state to true again, start a send, and return SUCCESS. Only one of the two packets will be sent successfully, but barring additional error checks in the call path, it can be hard to find out which one, and this might introduce all kinds of bugs in the calling components. Note that to avoid this problem, the send command isn't **async**.

> **Programming Hint 21** KEEP CODE SYNCHRONOUS WHEN YOU CAN. CODE SHOULD BE ASYNC ONLY IF ITS TIMING IS VERY IMPORTANT OR IF IT MIGHT BE USED BY SOMETHING WHOSE TIMING IS IMPORTANT.

11.1.3 Atomic statements and the atomic keyword

The problems interrupts introduce means that programs need a way to execute snippets of code that won't be preempted. nesC provides this functionality through atomic statements. For example:

```
command bool increment() {
  atomic {
    a++;
    b = a + 1;
  }
}
```

Listing 11.6 Incrementing with an atomic statement

The atomic block promises that these variables can be read and written atomically. In practice, on all current mote platforms, atomic statements are implemented by disabling and conditionally reenabling interrupts around the body of the atomic statement (this typically costs just a few cycles). In theory, a nesC implementation could allow incrementA and incrementC to preempt each other in this code snippet, as they access disjoint variables:

```
async command bool incrementA() {
  atomic {
    a++;
    b = a + 1;
  }
}
async command bool incrementC {
  atomic {
    c++;
    d = c + 1;
  }
}
```

Listing 11.7 Incrementing with two independent atomic statements

However, in practice implementing this is tricky, can have significant overhead and makes dealing with I/O devices (including a microcontroller's built-in peripherals) difficult.[1] Thus you should assume that atomic statements disable interrupts, and program in consequence (in particular, see the discussion on avoiding lengthy computation in atomic statements below).

nesC goes further than providing atomic statements: it also checks to see whether variables aren't protected properly and issues warnings when this is the case. For example, if b and c from the prior example didn't have atomic statements, then nesC would issue a warning because of possible self-preemption. The rule for when a variable has to be protected by an atomic statement is simple: if it is accessed from an async function, then it must be protected. nesC's analysis is sensitive to call sites: if you have a function that does not include an atomic statement, but is always called from within an atomic statement, the compiler won't issue a warning.

While you can make data race warnings go away by liberally sprinkling your code with atomic statements, you should do so carefully. On one hand, disabling and enabling interrupts around an atomic statement does have a CPU cost (a few cycles), so you want to minimize how many you have. On the other, shorter atomic statements delay interrupts less and so improve system concurrency. The question of how long an atomic statement runs is a tricky one, especially when your component has to call another component.

For example, the SPI bus implementation on the ATmega128 has a resource arbiter (Section 11.2) to manage access to the bus. The arbiter allows different clients to request the resource (the bus) and informs them when they've been granted it. However, the SPI implementation doesn't want to specify the arbiter policy (e.g. first come first served vs. priority), so it has to be wired to an arbiter. This decomposition has implications for power management. The SPI turns itself off when it has no users, but it can't know when that is without calling the arbiter (or replicating arbiter state). This means that the SPI has to atomically see if it's being used, and if not, turn itself off:

```
atomic {
  if (!call ArbiterInfo.inUse()) {
    stopSpi();
  }
}
```

In this case, the call to inUse is expected to be very short (in practice, it's probably reading a state variable). If someone wired an arbiter whose inUse command took 1ms, then this could be a problem. The implementation assumes this isn't the case. Sometimes (like this case), you have to make these assumptions, but it's good to make as few as possible.

---

[1] The interested reader may wish to read about software-transactional-memory implementation techniques.

11.1.4 Managing state transitions

The most basic use of atomic statements is for state transitions within a component. Usually, a state transition has two parts, both of which are determined by the existing state and the call: the first is changing to a new state, the second is taking some kind of action. Looking at the AMStandard example once again:

```
if (!state) {
   state = TRUE;
   // send a packet
   return SUCCESS;
}
else {
   return FAIL;
}
```

If state is touched by an async function, then you need to make the state transition atomic. But you don't want to make the entire block an atomic statement, as sending a packet could take a long enough time that it causes the system to miss an interrupt. So the code does something like this:

```
uint8_t oldState;
atomic {
   oldState = state;
   state = TRUE;
}
if (!oldState) {
   //send a packet
   return SUCCESS;
}
else {
   return FAIL;
}
```

If state were already true, it doesn't hurt to just set it true. This takes fewer CPU cycles than the somewhat redundant statement of

```
if (state != TRUE) {state = TRUE;}
```

In this example, the state transition occurs in the atomic block, but then the actual processing occurs outside it, based on the state the component started in.

11.1.5 Example: CC2420ControlP

Let's look at a real example. This component is CC2420ControlP, which is part of the CC2420 radio stack. CC2420ControlP is responsible for configuring the radio's various I/O options, as well as turning it on and off. Turning the CC2420 radio has four steps:

1. Turn on the voltage regulator (0.6 ms)
2. Acquire the SPI bus to the radio (time depends on contention)
3. Start the radio's oscillator by sending a command over the bus (0.86 ms)
4. Put the radio in RX mode (0.2 ms)

Some of the steps that take time are split-phase and have async completion events (particularly, 1 and 3). The actual call to start this series of events, however, is SplitControl.start, which is sync. One way to implement this series of steps is to assign each step a state and use a state variable to keep track of where you are. However, this turns out not to be necessary. Once the start sequence begins, it continues until it completes. So the only state variable you need is whether you're starting or not. After that point, every completion event is implicitly part of a state. For example, the startOscillatorDone event implicitly means that the radio is in state 3. Because SplitControl.start is sync, the state variable can be modified without any atomic statements:

```
command error_t SplitControl.start() {
  if ( m_state != S_STOPPED )
    return FAIL;

  m_state = S_STARTING;
  m_dsn = call Random.rand16();
  call CC2420Config.startVReg();
  return SUCCESS;
}
```

Listing 11.8 The first step of starting the CC2420 radio

The startVReg command starts the voltage regulator. This is an async command. In its completion event, the radio tries to acquire the SPI bus:

```
async event void CC2420Config.startVRegDone() {
  call Resource.request();
}
```

Listing 11.9 The handler that the first step of starting the CC2420 is complete

In the completion event (when it receives the bus), it sends a command to start the oscillator:

```
event void Resource.granted() {
  call CC2420Config.startOscillator();
}
```

Listing 11.10 The handler that the second step of starting the CC2420 is complete

Finally, when the oscillator completion event is signaled, the component tells the radio to enter RX mode and posts a task to signal the startDone event. It has to post a task because oscillatorDone is async, while startDone is sync. Note that the component also releases the bus for other users.

```
async event void CC2420Config.startOscillatorDone() {
  call SubControl.start();
  call CC2420Config.rxOn();
  call Resource.release();
  post startDone_task();
}
```

Listing 11.11 The handler that the third step of starting the CC2420 radio is complete

Finally, the task changes the radio's state from STARTING to STARTED:

```
task void startDone_task() {
  m_state = S_STARTED;
  signal SplitControl.startDone( SUCCESS );
}
```

Listing 11.12 State transition so components can send and receive packets

An alternative implementation could have been to put the following code in the startOscillatorDone event:

```
atomic {
  m_state = S_STARTED;
}
```

The only possible benefit in doing so is that the radio could theoretically accept requests earlier. But since components shouldn't be calling the radio until the startDone event is signaled, this would be a bit problematic. There's no chance of another task sneaking in between the change in state and signaling the event when both are done in the startDone_task.

Programming Hint 22 KEEP ATOMIC STATEMENTS SHORT, AND HAVE AS FEW OF THEM AS POSSIBLE. BE CAREFUL ABOUT CALLING OUT TO OTHER COMPONENTS FROM WITHIN AN ATOMIC STATEMENT.

11.1.6 Tasks, revisited

Chapter 5 introduced tasks as a way to defer computation and keep a nesC program responsive. They play an additional, critical role, however, with respect to asynchronous code. Tasks are the only way a component can transition from async to sync. Written as an interface, a task looks like this:

```
interface TaskBasic {
  async command error_t post();
  event void run();
}
```

All split-phase interfaces with an asynchronous command and a synchronous completion event must go through a task.

Note that the opposite direction – a sync command and async event – is rare. Such an interface would be for a service that can only be invoked from a task, but which requires the user to handle an interrupt. Because the event is async, any state shared between the command and event must be protected with atomic statements. Therefore, making the command async as well can enable a program to call it directly from the event, while typically not requiring any greater degree of code complexity.

11.2 Power locks

Atomic statements allow a component to execute a code block without interruption. However, low-level drivers often need to perform a series of split-phase operations on a single hardware resource without interference from any other part of the system. These drivers cannot use atomic statements – the completion of even the first operation occurs in a different function (the event handler) which clearly cannot be in the same atomic statement. Furthermore, disabling interrupts (as atomic statements normally do) for the length of time it takes to perform even one split-phase operation would probably be a bad idea.

In these situations, the atomicity is not on the processor – it's OK if other code executes – but rather on the hardware resource itself. To enable writing such drivers, TinyOS has *power locks*, which can be used to get exclusive access to a particular, typically hardware, resource. In addition to managing concurrency, power locks also manage energy and help configure hardware. While applications almost never see them, power locks constitute a key part of the internals of TinyOS and are important for most low-level systems work.

11.2.1 Example lock need: link-layer acknowledgements

Link-layer packet communication is one such example. TinyOS link-layer stacks support acknowledgements through the PacketAcknowledgements interface. When a node receives a packet destined to it, it immediately transmits a tiny packet in response, acknowledging reception. On the transmit side, the stack waits a short period for this link-layer acknowledgement, and reports whether it heard the acknowledgement. Minimizing the time the transmitter has to wait is critical for performance, so a receiver wants to send the acknowledgement as soon as possible.

Thinking through this problem from a programming standpoint, the receiver part of the radio stack needs to go through these steps:

- Read the packet out of the radio
- Inspect the packet to see if it should send an acknowledgement
- Switch the radio to transmit mode
- Send the acknowledgement
- Return the radio to receive mode

Each of these steps is typically a separate split-phase operation. On the CC2420 radio, for example, each operation requires sending a command that reads or writes data over an SPI bus. This SPI bus, however, is shared across many chips and subsystems: a flash storage driver might want to use the SPI bus to read data at the same time a node is receiving a packet. We can't rely on the SPI bus component to schedule multiple outstanding requests, as it might try interleaving a large flash write operation between two of the radio stack's. We need a way for the radio stack to request exclusive access to the bus so it can quickly perform its five operations, then release the bus for others to use.

There are numerous other cases where low-level systems need exclusive access to a hardware or software resource, either for correctness or latency; we'll cover a few examples later.

11.2.2 Split-phase locks

TinyOS supports exclusive access to resources through split-phase locks. Traditionally, locks such as mutexes and semaphores are blocking constructs that protect critical sections or shared data structures. However, as TinyOS does not have blocking calls, its locks must be split-phase. A component calls a command to request a power lock and receives an event when it acquires the lock. For historical reasons, the lock interface is named Resource:

```
interface Resource {
  async command error_t request();
  async command error_t immediateRequest();
  event void granted();
  async command void release();
  async command uint8_t getId();
}
```

Listing 11.13 The Resource interface

To acquire a lock, a component typically calls request. At some point later, it receives a granted event, signaling that it can use whatever the lock protects. The command immediateRequest is an optimization: it allows a component to acquire an idle lock in a single-phase operation. If immediateRequest returns FAIL then the lock is already held and the component must call request. Components release a lock with the release command.

Abstractions that require locking typically provide the Resource interface. Calling functional interfaces without holding the lock is typically forbidden, and may cause bugs or undesirable system behavior. For example, this is the abstraction nodes with an MSP430 microcontroller provide for an SPI bus:

```
generic configuration Msp430Spi0C() {
  provides interface Resource;
  provides interface SpiByte;
  provides interface SpiPacket;
  uses interface Msp430SpiConfigure;
}
```

Listing 11.14 Msp430Spi0C signature

Similarly, this is the abstraction for the ADC:

```
generic configuration Msp430Adc12ClientC() {
  provides {
    interface Resource;
    interface Msp430Adc12SingleChannel;
    interface Msp430Adc12MultiChannel;
    interface Msp430Adc12Overflow;
  }
}
```

Listing 11.15 Msp320Adc12ClientC signature

11.2.3 Lock internals

TinyOS *power locks* manage concurrency, energy, and configuration of shared hardware resources. Power locks have three sub-components:

- an *arbiter*, which controls the locking policy;
- a *power manager*, which controls the energy policy;
- and one or more *configurators*, which configure hardware for a client.

Figure 11.1 shows how these three parts fit together. The arbiter is the central point of control. An arbiter receives lock requests from clients, and calls the power manager and configurators based on those requests. Locks use parameterized interfaces and unique() to maintain a queue of pending lock requests. Each lock client has a unique lock ID, and the lock implementation uses uniqueCount() to maintain a queue of the proper length.

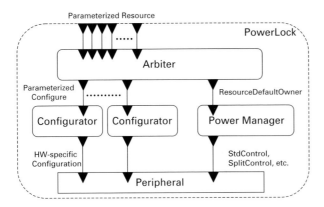

Figure 11.1 Power lock architecture.

11.2.4 Energy management

TinyOS uses locks for much more than just establishing exclusive access. It also uses locks to decide when to power peripherals on and off. Take the SPI bus as an example. Microcontrollers often have several low-power states, which differ in terms of which chip subsystems and interrupt sources are active. The lowest power state is typically one where a node can only wake up in response to one specific hardware counter or an external interrupt source: systems such as buses and ADCs are powered down and do not operate. Therefore, if TinyOS can disable the SPI bus when it's not in use, then the microcontroller can possibly save power.

Locks provide an easy way to automatically manage energy, lifting the burden from an application programmer. When a lock falls idle – it's released and there are no pending requests – it can power down the system it protects. When a lock becomes busy – it receives a request while in the idle state – it can power up the system. Depending on the nature of the underlying hardware, locks can either give a brief timeout before powering down or do so immediately.

Power locks support a "default owner," a component to which the lock reverts when it falls idle. This default owner is responsible for the power lock's energy management policy. A default owner uses a slightly different interface than Resource, as it never requests a lock. Instead, the arbiter always grants it an idle lock and notifies the default owner when there is a pending request. In response to a request, the default owner can power up the hardware and, once powered, release its lock:

```
interface ResourceDefaultOwner {
  async event void granted();
  async command error_t release();
  async command bool isOwner();
  async event void requested();
  async event void immediateRequested();
}
```

Listing 11.16 The ResourceDefaultOwner interface

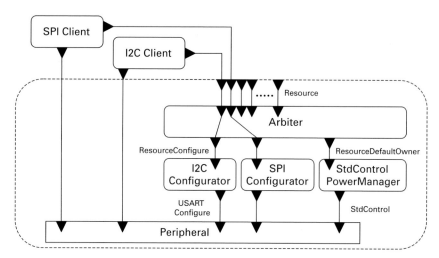

Figure 11.2 MSP430 USART stack. Instantiating an SPI client creates a wiring to the power lock's Resource interface, and couples that with a configuration interface. Before the arbiter grants the bus to the SPI client, it calls the SPI configuration code. The same is true of an I²C client.

11.2.5 Hardware configuration

In addition to controlling power states, power locks can configure hardware. Take, for example, the ADC on a microcontroller. ADCs typically have several configuration parameters, such as which pin's voltage to measure and what reference voltage to compare it to. While the power lock could just expect each client to do this configuration, it can also make the user's job easier by doing it automatically. When a configuration instantiates an ADC client, it passes its configuration parameters to the client component. These parameters are then passed to a configurator, which is wired in to the arbiter. Before the arbiter grants the lock to a client, it calls the corresponding configurator in order to set up the hardware as needed. When the client releases the lock, the arbiter unconfigures the hardware and returns to to a ready state:

```
interface ResourceConfigure {
  async command void configure();
  async command void unconfigure();
}
```

Listing 11.17 The ResourceConfigure interface

11.2.6 Example: MSP430 USART

The MSP430 USART0 (Universal Synchronous/Asynchronous Receiver/Transmitter) is a generic bus that software can configure for a variety of uses. The USART can be used as a UART (serial port), an SPI (Serial Peripheral Interface), or an I²C (Inter-Integrated

Circuit) bus. These three protocols share the same set of hardware pins, and so the software configuration tells the MSP430 how to clock them.

Software drivers use the bus protocols by instantiating a generic component, such as Msp430Spi0C, Msp430Uart0C, and Msp430I2COC. Each of these provides their functional interfaces and a Resource interface. Finally, they use a hardware-specific configuration interface. In the case of the SPI, for example, a client needs to configure the bus speed. This allows a client to place itself on the configurator path. Figure 11.2 shows this component structure.

11.2.7 Power lock library

TinyOS has a small library of reusable power lock components, which can be found in tos/system. They include:

- Two arbiters with different scheduling policies: FcfsArbiterC (first-come, first-served) and RoundRobinArbiterC (round-robin)
- Two power managers: ImmediatePowerManagerC and DeferredPowerManagerC
- Numerous configurators: as configurators are hardware-specific, they are spread throughout the code base, and are mostly in tos/chips

If you want more detailed information on power locks, their performance, and use, please refer to TEPs 108 [13] and 115 [11], and to the paper describing them [12].

11.3 Exercises

1. The MSP430 ADC has a deferred power manager. Its timeout used to be 100 ms, but after some experiments it was changed to a timeout of 20 ms. Why 20 ms? Hint: look at how long it takes for the voltage reference to stabilize.
2. Why don't radios typically have a power lock?
3. Compute the CPU cost of an atomic statement. Write a nested atomic section, where there is one atomic statement within another. Does the overhead go up?
4. Most interrupt handlers are themselves atomic: they are completely within an atomic statement. Try writing a UART byte reception interrupt handler that is re-entrant (can re-execute while already executing). Why is it hard? Can you do it without any atomic statements at all?

12 Device drivers and the hardware abstraction architecture (HAA)

By their very nature, sensor network applications are often platform-specific: the application uses a particular set of sensors on a mote-specific sensor board, to measure application-specific conditions. The hardware, and hence the application, is typically not directly reusable on another mote platform. Furthermore, some applications may want to push a mote platform to its limits, e.g. to maximize sampling rate, or minimize the latency in reacting to an external event. Getting to these limits normally requires extensive platform-specific tuning, including platform-specific code (possibly even written in assembly language).

Conversely, large portions of sensor network applications *are* portable: multi-hop network protocols, radio stacks for commonly available radio chips, signal processing, etc. need little or no change for a new platform. Thus, while a sensor network application is not typically directly portable, TinyOS *should* make it easy to port applications to new platforms by minimizing the extent of the necessary changes.

12.1 Portability and the hardware abstraction architecture

TinyOS's main tool to maximize portability while maintaining easy access to platform-specific features is a multi-level hardware abstraction architecture (HAA), shown in Figure 12.1. The device driver components that give access to a mote's various hardware resources are divided into three categories:

- The hardware interface layer (HIL): a device driver is part of the HIL if it provides access to a device (radio, storage, timers, etc.) in a platform-independent way, using only hardware independent interfaces.
  The functionality of the HIL is limited by what is available on multiple platforms. For instance, while some flash chips incorporate some form of write protection, this functionality is not common to all flash chips, so not reflected in the storage layer HIL (Section 6.5).
  Except for the sensors, all the components (for communication, storage, etc.) we used for the anti-theft demo in Chapter 6 are part of the HIL.
- The hardware adaptation layer (HAL): device drivers in the HAL simplify the use of the often complex underlying hardware, by exposing it via high-level interfaces. For instance, the HAL for the Atmel AT45DB-family of flash chips provides operations to

12.1 Portability and the HAA

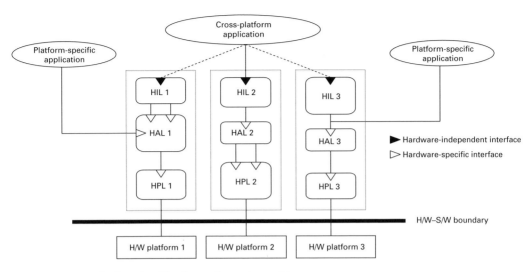

Figure 12.1 TinyOS's three-level hardware abstraction architecture.

read and write parts of the flash's storage blocks, while automatically managing the flash chip's 2 RAM buffers.

HAL components are platform-specific, but should use hardware independent interfaces when possible. For instance, the Alarm interface used internally in the micaz's HilTimerMilliC component (Section 9.2.1) is provided by a micaz-specific AlarmCounterMilliP component, but used by the portable AlarmToTimerC component that implements a Timer interface given an Alarm interface.

- The hardware presentation layer (HPL): device drivers in the HPL sit directly above the hardware without abstracting away any of its functionality. The goal of HPL components is to hide irrelevant differences between similar hardware and to present hardware functionality in a nesC-friendly fashion.

For instance, on the micaz platform, the HplAtm128GeneralIOC exposes the ATmega128's 53 digital I/O pins as 53 GeneralIO interfaces, hiding the slightly different instruction sequences needed to perform some operations on some I/O pins (some I/O pins can be set atomically in a single assembly instruction, while others require interrupts to be disabled to guarantee atomicity).

As shown in Figure 12.1, a device's HIL is normally built on top of the device's HAL, which is itself built on top of the HPL.

The behaviour, name and specifications of TinyOS's hardware independent layers are specified in TinyOS Enhancement Proposals (TEPs), and the hardware abstraction architecture itself is described in TEP 2 [8]. TEPs do not specify the contents of the HAL or HPL for any hardware, but do sometimes provide guidelines on HAL structure, and on the use of hardware independent interfaces in the HAL and HPL.

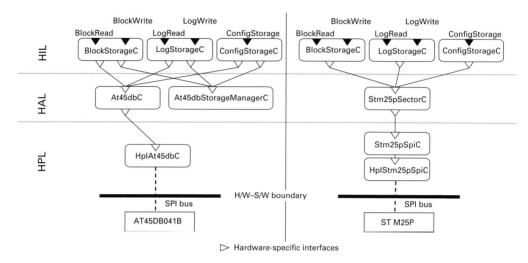

Figure 12.2 HPL, HAL, and HIL Storage Components in TinyOS.

12.1.1 Examples

Figure 12.2 illustrates a typical example of the hardware abstraction architecture: the structure of the storage for two different flash chips, Atmel's AT45DB041B serial data flash, and ST's M25P serial NOR flash. Both chips support the three HIL storage abstractions that we saw in Section 6.5, BlockStorageC, LogStorageC, and ConfigStorageC. However, the implementations of these abstractions is different for each chip. The HAL layers for each chip are composed of different numbers of components, and provide different interfaces:

```
configuration At45dbC {
  provides interface At45db;
  ...
}
module At45dbStorageManagerC {
  provides interface At45dbVolume[volume_id_t volid];
}
configuration Stm25pSectorC {
  provides interface Stm25pSector as Sector[uint8_t id];
  ...
}
```

The AT45DB041B HAL is provided by two components: At45dbC provides a single interface (At45db) with erase, read, write, and CRC-computation commands, and manages a small on-flash-chip cache. The At45dbStorageManagerC component returns the configuration of a specific TinyOS storage volume given its id. In contrast, the ST M25P HAL is provided by a single component, Stm25pSectorC, which combines high-level erase, read, write, and CRC computation (in the Stm25pSector interface) with volume management. Stm25pSectorC provides a parameterized interface where the interface parameter identifies each individual HAL user.

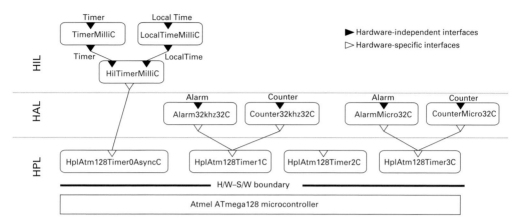

Figure 12.3 HPL, HAL, and HIL timer components on the micaz mote.

The two chips, and hence the two storage HPLs have one significant similarity: both chips are accessed via an SPI bus. As a result, each platform must provide glue code that connects the platform-independent (but chip-dependent) HPL code to a platform-specific SPI bus implementation. For the ST M25P, this connection is specified by the platform-specific HplStm25pSpiC component, and Stm25pSpiC contains platform-independent code with low-level erase, read, write, and CRC computation operations. For the AT45DB041B, each platform must implement the HplAt45dbC component which provides low-level erase, read, write, CRC computation, status, and cache-management operations.

HAA structures vary significantly across devices. For instance, Figure 12.3 summarizes TinyOS's support for timers on the micaz platform. At the lowest layer, four HPL components provide access to the four ATmega128 timers. Two of these timers (0 and 2) are 8-bit, and two (1 and 3) are 16-bit. Furthermore, timer 0 supports an "asynchronous" mode where it is clocked by an external (usually 32 768 Hz) crystal, while the other timers are clocked from the same source as the microcontroller, an external 7.37 MHz crystal on the micaz. On the mica, timer 0 is dedicated to implementing TinyOS's timer subsystem, HilTimerMilliC:

```
configuration HilTimerMilliC {
  provides interface Init;
  provides interface Timer<TMilli> as TimerMilli[uint8_t num];
  provides interface LocalTime<TMilli>;
}
```

The user-level timer components, TimerMilliC (and LocalTimeMilliC) are simple portable wrappers over HilTimerMilliC (see Section 9.2.1). At the HAL level, the micaz provides 32-bit counters and alarms using 1/32768 s and μs time units. The Alarm and Counter interfaces provided by these components are hardware independent, as are the component names (Alarm32khz32C, etc.). However, these components need not be provided by all platforms, and the number of available alarms – three in the case of the micaz – varies from platform to platform. As a result, these components belong to the HAL rather than the HIL. However, the use of the same names and interfaces across platforms does simplify writing reasonably portable code with timing requirements that cannot be satisfied by the fully portable TimerMilliC component. Finally, as the figure shows, timers 1 and 3 are dedicated to supporting the micaz's HAL layer, but timer 2 is unused and available to an application that needs it.

12.1.2 Portability

While a program or component that uses only device drivers that are part of the HIL could be viewed as being "fully" portable, the reality is more complex. Different motes have different hardware resources, and hence end up supporting different HILs: for instance, a hypothetical "Ghost" mote without a flash chip or other form of permanent storage will not support the log storage abstraction of Section 6.5. Therefore, the FlashSampler application from Section 6.5.3 would not be portable to "Ghost" motes.

In contrast, the anti-theft application from the first part of Chapter 6 depends on a light sensor component (PhotoC) and an accelerometer component (AccelXStreamC) which are not part of the HIL – these two components are specific to a particular sensor board (mts310) for a particular platform (micaz). However, the anti-theft code accesses these components via hardware independent interfaces (Read and ReadStream respectively), so is fairly easy to port to a different mote as long as it provides equivalent sensors: only the sensor component names and detection thresholds (DARK_THRESHOLD and ACCEL_VARIANCE) should need changing.

In summary, applications that use only HIL components, or access HAL components via hardware-independent interfaces should be easily portable to motes which meet the application's hardware requirements. The more the application accesses HAL or HPL components, especially via hardware-specific interfaces, the harder it becomes to port.

12.2 Device drivers

A TinyOS program can choose to access a particular hardware device via its HPL, HAL, and HIL components. However, not all devices have or need HIL components: if a device exists on only one platform or sensor board and does not much resemble any other device, then it is unlikely that anyone will have written a device-independent HIL specification. Additionally, some hardware does not really offer any hardware-independent interface. For instance, the behavior and configuration of an A/D converter is intimately tied to the specific analogue sensor which is connected to the converter. As a result, there is no real hardware-independent A/D interface, and hence no A/D HIL.

The choice between HPL, HAL, and HIL is at one level a choice between ease of use, portability, functionality, and performance. The HPL components, which expose the raw hardware, provide, of course, the best functionality and performance. The HIL components are the most portable, and typically the easiest to use. The HAL components are usually a compromise between these two extremes. However, beyond differences in functionality, the different hardware abstraction layers also differ in how they handle multiple clients, i.e. perform access control, and in how they manage power usage. In the rest of this chapter, we examine these last two issues in greater detail.

12.2.1 Access control

Device drivers need to perform access control to prevent problems when multiple parts of a program try to access the same device (because TinyOS only runs a single application, access control across applications is not an issue). For instance, the ATmega128 A/D converter can only sample one channel at a time, so in a program where the radio needs to check signal strength on the antenna and the user's code needs to sample the current temperature, some coordination will be needed to avoid the radio and the user's code corrupting each other's requests.

TinyOS 1.x mostly relied on detecting conflicts and programmer discipline to deal with access control issues. For instance, a request to perform A/D conversion would return an error if some other part of the program was already using the A/D converter:

```
/* Let's read temperature! */
if (call ADC.getData() == FAIL)
  /* oops, we need to try again */
  post tryAgain();
```

In practice, this didn't work. Programmers omitted error checking, or could not figure out good error recovery strategies. Furthermore, these error paths were unlikely (e.g. would only occur if the radio and application actually tried to sample at the same time), so were not well tested. Finally, the resulting code is often complex, so is hard to get right, understand, and maintain. Evolution of the rest of the system makes life even more complicated: user code that samples temperature without checking the error codes will, e.g., work fine with a radio stack that does not sample signal strength. However, if the radio stack is changed to one that does sample signal strength, then the user's application code may suddenly stop working.

As a result, TinyOS 2 takes a much more systematic approach to access control. All device driver components are expected to fit in to one of three access control classes:

- *Dedicated* A dedicated driver has a single user, who has full control over device. Examples include most HPL components.
- *Virtualized* A virtualized driver has multiple users, each of which appears to have its own instance of the device. However, the underlying implementation creates this illusion by multiplexing a single underlying resource. For instance, each instance of a TimerMilliC component is an independent timer, but they are all implemented over a single hardware timer, as discussed in Section 9.2.1.
- *Shared* A shared driver has multiple users, who coordinate their use of the device via the the Resource interface (Section 11.2). The HAL components for both of the storage chips we saw above (Atmel AT45DB041B and ST M25P) are shared drivers.

The definitions of these three classes uses the term "user" to informally refer to entities like the radio stack, the user's application code, etc. In practice, a "user" of a driver is ultimately some nesC module which is wired to some component or interface representing an instance of the driver.

12.2.2 Access control examples

A dedicated driver has a single instance, normally a component with a specific name. HPL drivers are normally dedicated; examples include HplAtm128Timer2C (HPL for timer 2 of the ATmega128 microcontroller), HplMsp430I2C0C (HPL for the I^{2}C bus 0 on the TI MSP430 microcontroller). For example, in the following excerpt:

```
module HplAtm128Timer2C
{
  provides interface HplAtm128Timer<uint8_t> as Timer;
  provides interface HplAtm128TimerCtrl8 as TimerCtrl;
  provides interface HplAtm128Compare<uint8_t> as Compare;
} implementation {
  async command uint8_t Timer.get() { return TCNT2; }
  async command void Timer.set(uint8_t t) { TCNT2 = t; }
  ...
}
```

we see that HplAtm128Timer2C provides only functional interfaces (Timer, TimerCtrl, Compare) which are very simple wrappers over the raw hardware (the TCNT2 timer 2 counter register in this excerpt). Because there are no issues with sharing the resource with other users, dedicated drivers offer the lowest performance and latency overhead — operations are performed immediately and with no checking.

Nothing prevents multiple users from wiring to HplAtm128Timer2C and using timer 2 in a conflicting fashion. An HPL implementer can prevent such multiple wirings using nesC's @atmostonce() wiring attribute (Section 8.4), e.g.:

```
provides interface HplAtm128Timer<uint8_t> as Timer @atmostonce();
```

However, using these attributes is currently uncommon as it sometimes causes complications: with the change above, it would no longer be possible to split code using timer 2 into two separate (but tightly cooperating) modules that both wire to HplAtm128Timer2C.

Dedicated drivers are not always low-level components. For instance, the ActiveMessageC radio link layer abstraction is a dedicated driver with the following signature:

```
configuration ActiveMessageC {
  provides {
    interface SplitControl;
    interface AMSend[uint8_t id];
    interface Receive[uint8_t id];
    interface Receive as Snoop[uint8_t id];
    interface Packet;
```

```
    interface AMPacket;
    interface PacketAcknowledgements;
  }
}
```

Listing 12.1 ActiveMessageC signature

ActiveMessageC provides a means of powering the radio on and off (see the discussion of power management below) and interfaces to send, receive, and inspect packets. However, it does not provide an easy way for multiple users to cooperate in sending packets via the AMSend interface – like in TinyOS 1.x, if the radio is currently sending a packet, a second send command will simply fail with an error result (`EBUSY`):

```
/**
 * Send a packet
 * @return     ... EBUSY if the abstraction cannot send now but
 *                 will be able to later
 */
command error_t send(am_addr_t addr, message_t* msg, uint8_t len);
```

The AMSenderC component we saw in Chapter 6 is a virtualized driver built over ActiveMessageC and, like most virtualized drivers, is a generic component:

```
generic configuration AMSenderC(am_id_t AMId) {
  provides {
    interface AMSend;
    interface Packet;
    interface AMPacket;
    interface PacketAcknowledgements as Acks;
  }
}
```

AMSenderC offers the same interfaces to send and inspect packets as ActiveMessageC, though the `id` parameter to the AMSend interface has become AMSenderC's `AMId` generic component parameter. Instances of such virtualized drivers are simply created by instantiating the generic component:

```
components new AMSenderC(AM_THEFT) as SendTheft;
MovingC.AMSend -> SendTheft;
...
```

The practical difference between AMSenderC and ActiveMessageC is that each instance of AMSenderC can have one outstanding packet, where ActiveMessageC allowed only one outstanding packet for the whole system – AMSenderC is virtualized while ActiveMessageC is dedicated. In general, virtualized drivers are a lot easier to use as you can ensure that your code always has a resource (a packet transmission slot, a timer, etc.) available when it requires one, rather than having to deal with sharing the resource with the rest of the system. As a result, most HIL drivers are virtualized. However, virtualized drivers have two costs. First, managing and selecting between the requests from all users adds some run-time overhead. Second, the latency of any individual operation is unpredictable: some arbitrary set of operations from other users may happen between your request and its actual execution. Some virtualized drivers may however offer some specific guarantees, e.g. the AMSenderC instances handle their users in a round-robin fashion.

Shared drivers offer a workaround to the unpredictable latency of virtualized drivers. Shared drivers use the Resource interface (Section 11.2) to control access by multiple users, and once a user has been granted access to the driver, that user's operations are executed immediately. Each user is identified by a unique id, obtained using nesC's unique() function (Section 9.1). This id is then used to parameterize both the Resource and functional interfaces, to easily distinguish each user's commands and events, as in Stm25pSectorC, the HAL driver for the ST M25P flash chip:

```
configuration Stm25pSectorC {
  provides interface Resource as ClientResource[uint8_t id];
  provides interface Stm25pSector as Sector[uint8_t id];
  provides interface Stm25pVolume as Volume[uint8_t id];
}
```

Users of Stm25pSectorC must request access via the ClientResource interface before making any requests to read, write, or erase the flash via Stm25pSector:

```
char mybuf[16];

void readMybuf() {
  /* Request access before starting the read */
  call FlashResource.request();
}

event void FlashResource.granted() {
  /* We have access, we can now do our read */
  call Stm25pSector.read(0, mybuf, sizeof mybuf);
}

event void Stm25pSector.readDone(...) {
  /* Release the resource now that we're done */
  call FlashResource.release();
  ... use mybuf ...
}
```

More complex code might perform multiple operations using Stm25pSector before releasing access to the flash. Note, however, that performing many long operations will increase the latency before which other flash users can perform any operations at all. As compared to virtualized drivers, shared drivers have predictable latency for operations performed when access has been granted. The cost is extra code complexity to manage the explicit driver request and release operations, and a possibly longer latency until the first operation can be executed.

Wiring to a shared driver such as Stm25pSectorC is generally straightforward. A client id is picked using unique(), and the functional and Resource interfaces are then wired to user code:

```
components MyAppC, Stm25pSectorC;
enum { MYID = unique(UQ_STM25P_VOLUME) };
MyAppC.FlashResource -> Stm25pSectorC.ClientResource[MYID];
MyAppC.Stm25pSectorC -> Stm25pSectorC.Stm25pSector[MYID];
```

Some shared drivers (e.g. Msp430I2CC, the TI MSP430's HAL for the I^2C bus) wrap this wiring logic into a generic component:

```
generic configuration Msp430I2CC() {
  provides interface Resource;
  provides interface I2CPacket<TI2CBasicAddr> as I2CBasicAddr;
  uses interface Msp430I2CConfigure;
}
```

Implementing access control for shared drivers is simplified by using TinyOS's power-lock library (Section 11.2.7). An arbiter provides an implementation of a parameterized Resource interface with a specific queuing policy, and a hook for power management (see below). The generic arbiter components take the string used to pick unique client ids as a parameter. For instance, the following line creates Stm25pSectorC's arbiter:

```
components new FcfsArbiterC(UQ_STM25P_VOLUME) as ArbiterC;
```

The implementation of Stm25pSectorC needs to perform a little work every time access is granted by the arbiter and released by the user, so interposes a module (Stm25pSectorP) between the ClientResource interface and its first-come, first-served arbiter (FcfsArbiterC):

```
configuration Stm25pSectorC {
  provides interface Resource as ClientResource[uint8_t id];
  ...
} implementation {
  components Stm25pSectorP as SectorP;
  components new FcfsArbiterC(UQ_STM25P_VOLUME) as ArbiterC;

  ClientResource = SectorP.ClientResource;
  SectorP.Stm25pResource -> ArbiterC.Resource;
  ...
}
```

Listing 12.2 Arbitration in Stm25pSectorC

12.2.3 Power management

Motes are usually powered by low-capacity batteries, so effective power management is essential for obtaining reasonable sensor network lifetime. For instance, to last a year, a mote powered by two 2700 mAh batteries (e.g. traditional AA batteries) must have an average power consumption of 0.9 mW. A typical microcontroller uses at least several mW when active, and a typical radio tens of mW when listening for messages. Luckily, radios can be switched off, and microcontrollers have various sleep modes where power consumption drops to levels as lows as a µw. Thus, to achieve an average of 0.9 mW, the mote must spend most of its time with the radio off and the microcontroller in some appropriate sleep state; the same reasoning applies to any other peripheral (sensor, flash chip, etc.) with non-trivial power consumption.

The basic decision as to when various components must be on rests in the application's hands: only it knows when sensors must be sampled, messages transmitted, etc. The goal of TinyOS's power management is "simply" to put every subsystem into the lowest power state consistent with these application demands. Power states vary from device to device, however, for current mote hardware, power states are usually:

- On and off for most peripherals, be they on-microcontroller like A/D converters, I^2C buses or off-microcontroller like sensors, flash chips.
- Radio-specific states, as in low-power-listening which listens for radio messages at some frequency, or schedule-based where motes transmit and receive at specific times. These states effectively tradeoff increased latency and reduced bandwidth for lower power usage.
- A hierarchy of power states for microcontrollers. For instance, an Atmel ATmega128 can be in active, idle, ADC-noise-reduction, extended standby, power save, standby, and power down states. Each successive state turns off more on-chip functionality and increases wakeup latency.

In TinyOS, only the selection of the radio state is left to the application programmer. Peripherals are powered on and off in most device drivers (except dedicated drivers) based on the pending user requests, while the microcontroller power state is derived from the whole system state.

Thus, the extent of user-level power management code in a program like the anti-theft application of Chapter 6 is the call to set the duty cycle after the radio has been started (and the associated wiring):

```
uses interface LowPowerListening;
...
event void CommControl.startDone() {
  // Switch radio to low-power-listening with a 2% duty cycle
  call LowPowerListening.setLocalDutyCycle(200);
  call TheftTimer.startPeriodic(ACCEL_INTERVAL);
}
```

Implementation of power management in device drivers depends heavily on the driver's class (dedicated, shared, virtualized). Dedicated drivers typically do not manage power directly, but offer commands to turn the device on and off for use by the dedicated driver's single user. With the exception of the radio stack (ActiveMessageC), dedicated drivers are normally managed by other device drivers, so this level of explicit power management is invisible to applications.

Most, but not all, dedicated drivers use one of the StdControl, SplitControl, or AsyncStdControl interfaces to control power. One exception is the HPL for the ATmega128 A/D converter, which offers the following two commands to switch the

A/D on and off:

```
async command void HplAtm128Adc.enableAdc() {
  SET_BIT(ADCSRA, ADEN);
}

async command void HplAtm128Adc.disableAdc() {
  CLR_BIT(ADCSRA, ADEN);
}
```

Power management is typically performed automatically by shared drivers. If the shared driver is implemented using a power lock, then the driver can instantiate a power manager (Section 11.2) to automatically switch itself on and off depending on its clients requests. The device must choose between the immediate and deferred power managers – in the deferred case, the system waits for a small time after the last user is done before switching the device off. A deferred power off avoids the cost of switching the device off and immediately back on if a new request comes in shortly (but not immediately) after the last one completed. Deferred is a good policy when switching a device on and off is a complex operation, when the device requires a "warm up" period before first use, etc. For instance, TinyOS's Stm25pSectorC implementation uses a deferred power manager (instantiated as PowerManagerC below) with a 1 s delay because powering the flash chip on or off requires sending a command to the flash via an SPI bus:

```
configuration Stm25pSectorC {
  provides interface Resource as ClientResource[uint8_t id];
  ...
} implementation {
  components Stm25pSectorP as SectorP;
  components new FcfsArbiterC(UQ_STM25P_VOLUME) as ArbiterC;
  ClientResource = SectorP.ClientResource;
  SectorP.Stm25pResource -> ArbiterC.Resource;

  // 1024 binary milliseconds = 1 second power off delay:
  components new SplitControlDeferredPowerManagerC(1024) as PowerManagerC;
  PowerManagerC.SplitControl -> SectorP;
  PowerManagerC.ResourceDefaultOwner -> ArbiterC;
  PowerManagerC.ArbiterInfo -> ArbiterC;
  ...
}
```

Virtualized devices are most often built over shared devices that already perform power management, so the virtualized device contains no explicit power management code. However, if a virtualized device does need to manage power, the situation is similar to the arbiter case: the virtualized device knows all the outstanding requests for the underlying hardware, so can can decide when to power the device on and off.

12.2.4 Microcontroller power management

As we discussed above, microcontrollers can typically enter one of several different sleep modes when no code needs to run. Execution resumes on the next hardware interrupt. The details of the sleep modes and how to enter them are very device-specific, but the decision of when to sleep is very simple in TinyOS: when the task queue is empty, the mote can sleep.

The sleep logic is encapsulated in a microcontroller-specific McuSleepC component:

```
module McuSleepC {
  provides interface McuSleep;
  provides interface McuPowerState;
  uses interface McuPowerOverride;
}
implementation { ... }
```

Listing 12.3 McuSleepC: platform-specific sleep code

The McuSleep interface has a single command, sleep, that tells McuSleepC to put the microcontroller in the lowest-power-consumption sleep mode consistent with the mote's current configuration. For instance, on an ATmega128, the microcontroller must stay in the highest power-consumption mode ("idle") if the serial ports, SPI or I^{2}C bus are in use. Conversely, if the mote is only waiting for a timer to expire, then the ATmega128 can be put in the "power-save" mode, dropping microcontroller power consumption from tens of mW to tens of μW. There is, however, a cost to entering power-save mode: when using, e.g., an external crystal oscillator, the microcontroller takes 16 000 cycles to wake up from power-save mode. At 8 MHz, 16 000 cycles is 2 ms, so power-save should not be entered if the timer will expire in the next few ms.

The McuSleep.sleep command for the ATmega128 implements the logic outlined above (and more) by inspecting the microcontroller's hardware registers to find out which on-board peripherals (buses, timers, etc.) are enabled and pick the appropriate sleep mode. However, the selection of sleep mode is often dependent on some platform-specific features and even in some cases on application requirements (e.g. the delay in responding to a timer when in power-save mode may be problematic in an application that has high timing precision requirements). To allow for such variations, McuSleepC calls the McuPowerOverride.lowestState command to allow the rest of the system to force the choice of an appropriate power state. For instance, on the ATmega128 :

```
async command void McuSleep.sleep() {
  uint8_t powerState;

  powerState = mcombine(getPowerState(), call McuPowerOverride.lowestState());
  ... set sleep mode to powerState and go to sleep ...
}
```

```
default async command mcu_power_t McuPowerOverride.lowestState() {
  return ATM128_POWER_DOWN;
}
```

The logic in McuSleepC assumes that sleep modes are ordered by the mcombine function: for any two sleep modes s_1 and s_2, mcombine(s_1, s_2) is the best sleep mode which supports the functionality of both s_1 and s_2. On a microcontroller like the ATmega128 the powered modes are ordered from most-power/most-functionality ("idle") to least-power/least-functionality ("power-down"), so mcombine(s_1, s_2) is simply max(s_1, s_2).

To allow several components to override the sleep mode, McuPowerOverride supports fan-out wiring by defining mcombine as the combine function (Section 4.4.3) for McuPowerOverride.lowestState's result:

```
typedef uint8_t mcu_power_t @combine("mcombine");
```

Computing the best sleep mode has the potential to significantly affect a mote's lifetime: while the sleep mode computation is normally simple and therefore reasonably cheap, it is executed after nearly every interrupt. However, on many platforms, the result of this computation only changes as the result of powering some device on or off. As a result, to compute the sleep mode only when necessary, McuSleepC provides the McuPowerState.update command. Components (normally found in the HPL) that change the mote's state in a way that affects the sleep mode must call McuPowerState.update so that McuSleepC knows it needs to recompute the sleep mode. For instance, on the TI MSP430, the code is as follows:

```
bool dirty = TRUE;

async command void McuSleep.sleep() {
  if (dirty) {
    computePowerState();
    dirty = 0;
  }
  ... set sleep mode and sleep ...
}

async command void McuPowerState.update() {
  atomic dirty = 1;
}
```

12.3 Fitting in to the HAA

The Hardware Abstraction Architecture provides a general framework for building and using TinyOS services. The HIL defines functionality that all implementations provide, filling the role that standard APIs do in most systems. As some applications need to take advantage of hardware features, the HAL provides access to richer functionality in a way that makes it clear it is hardware-specific.

The HAA emerged from earlier versions of TinyOS due to observed incompatibilities between platforms. For example, platforms sometimes provided different signatures for a component with the same name, preventing cross-compilation. Writing cross-platform code therefore required looking through all of the platforms and figuring out the maximal subset that all implementations shared, with the understanding that a new platform might change it.

While we've presented the HAA as a clear three-layer hierarchy, in practice it tends to be a bit more complex, and at times even a point of disagreement within the TinyOS community! For example, platforms define the component ActiveMessageC to specify the default link-layer communication stack. ActiveMessageC is part of the HIL. However, radio chips also provide chip-specific abstractions, such as CC2420ActiveMessageC, and CC1000ActiveMessageC. These components provide all of the functionality of ActiveMessageC, as well as chip-specific operations. For example, CC2420ActiveMessageC has interfaces for accessing CC2420-specific values and fields. These chip-specific communication components have the same datapath interfaces as the HIL, and a few extra control interfaces. Whether these components lie in the HAL, the HIL, or a hazy place in-between is a source of much debate.

Rather than a set of hard rules, the HAA is a way to organize and name components that (hopefully) simplifies application development. It makes it clear when a component wires to a hardware-specific service, and also provides guidelines on what a component must provide to be cross-platform.

13 Advanced application: SoundLocalizer

In this chapter, we look at the design and implementation of SoundLocalizer, a somewhat more complex sensor network application. SoundLocalizer implements a coordinated event detection system where a group of motes detect a particular event – a loud sound – and then communicate amongst themselves to figure out which mote detected the event first and is therefore presumed closest to where the event occurs. To ensure timely event detection, and accurately compare event detection times, this application needs to use some low-level interfaces from the platform's hardware abstraction and hardware presentation layers (HAL, HPL, as described in the previous chapter). As a result, this application is not directly portable – we implement it here for micaz motes with an mts300 sensor board. In the design and implementation descriptions below, we discuss how the application and code are designed to simplify portability and briefly describe what would be involved in porting this application to another platform.

The HAL and HPL components used by SoundLocalizer offer lower-level interfaces (interrupt-driven, controlled by a Resource interface, etc.) than the high-level HIL components we used to build the AntiTheft application of Chapter 6. As a result, SoundLocalizer's implementation must use atomic statements and arbitration to prevent concurrency-induced problems, as we saw in Chapter 11.

The complete code for SoundLocalizer is available from TinyOS's contributed code directory (under "TinyOS Programming").

13.1 SoundLocalizer design

Figure 13.1 shows a typical setup for the SoundLocalizer application. A number of detector motes are placed on a surface a couple of feet apart. When the single coordinator mote is switched on, it sends a series of radio packets that let the detector motes synchronize their clocks. At a time specified by the coordinator mote, all detectors turn on their green LED and start listening for a loud sound such as a hand clap. Once such a sound is heard, the motes turn on their yellow LED. Finally, the motes enter a "voting" phase where only the mote with the earliest detection time leaves its yellow LED on. The earliest detection time should correspond to the mote closest to the sound, but various factors can invalidate this hypothesis, as we will discuss after explaining the application's design in more detail.

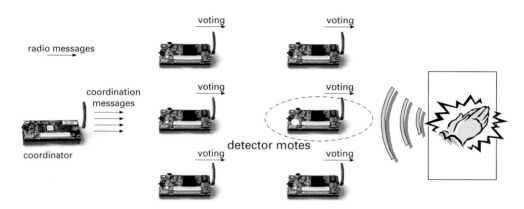

Figure 13.1 SoundLocalizer application setup.

We deliberately kept SoundLocalizer's design and implementation simple. All design choices were driven by a desire for clarity of exposition. A "real" implementation of a coordinated event detection system would require more engineering than shown here, to improve precision and reliability. However, SoundLocalizer's structure, and the way it accesses low-level platform-specific features is representative of what real applications do.

13.1.1 Time synchronization

SoundLocalizer requires motes to have a consistent view of time so that they can accurately compare their detection times. For instance, if two motes disagree by 10 ms on the current time, then one mote could be 3 meters further from a sound and still think it detected the sound first, as sound travels 3.4 meters in 10 ms. Today, TinyOS does not offer any time synchronization facility (though this is an area of active development), so SoundLocalizer includes its own simple time synchronizer. To simplify implementation, we use a dedicated coordinator mote that broadcasts a time signal, and assume that all detector motes are within radio range of the coordinator.

If all mote's clocks ran at precisely the same frequency, then time synchronization could be accomplished by having all detector motes d note the precise local arrival time t_d of a "synchronize" radio message sent by the coordinator mote: for practical purposes, the differences in radio propagation time from the coordinator mote to the various receivers are negligible. Given this time t_d, local time t could then be converted to a globally agreed time $t - t_d$: the reception time of the "synchronize" message is global time 0.

However, in practice each mote has a slightly different clock rate and may record slightly delayed arrival times because of other activities on the mote. To account for these differences, the coordinator sends synchronization messages containing the numbers $0, \ldots, N-1$ at fixed intervals. A mote d records the local arrival times $t_{d,i}$ of the message containing sequence number i, and uses least-squares linear regression to compute the

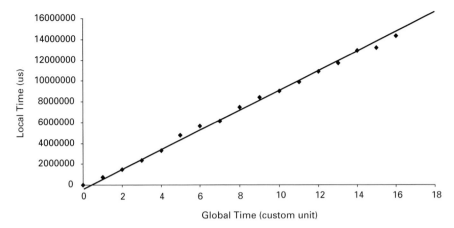

Figure 13.2 Linear Regression for Time Synchronization. The line represents the best estimate of the relation between local and global time.

best relation between local time and the "global time" created by the coordinator's synchronization messages (Figure 13.2).

The resulting time synchronization is sufficiently precise for micaz motes placed within a foot of each other to distinguish the arrival time of a loud sound. The motes can use either the micaz's external crystal oscillator or its internal RC oscillator (though in the case of the internal oscillator it is necessary to wait a few minutes after switching a mote on for the oscillator to stabilize).

More elaborate time synchronization algorithms [1, 22] can achieve significantly greater precision, avoid the need for a dedicated coordinator mote and handle multihop networks. However, these algorithms are more complex to implement.

13.1.2 Implementing SoundLocalizer in TinyOS

SoundLocalizer has separate programs for the detector and coordinator motes. The coordinator code is very simple: when the coordinator mote boots it starts a periodic timer:

```
call Timer.startPeriodic(BEACON_INTERVAL);
```

At each timer event, the coordinator mote sends a synchronization message with the next sequence number and a global time at which event detection is to occur (error checking is elided below):

```
typedef nx_struct coordination_msg {
  nx_uint16_t count, sample_at, interval;
} coordination_msg_t;

event void Timer.fired() {
  coordination_msg_t *payload=call AMSend.getPayload(&msg, sizeof *payload);
```

```
/* Send the current sequence number */
payload->count = count;
payload->sample_at = SAMPLE_TIME;
payload->interval = BEACON_INTERVAL;
call AMSend.send(AM_BROADCAST_ADDR, &msg, sizeof *payload);
/* Stop sending once the event detection time approaches */
if (++count == SAMPLE_TIME)
  call Timer.stop();
}
```

The detector code is significantly more complex. It is contained in five components (Figure 13.3):

- The SynchronizerC module contains the time synchronization algorithm, voting code and the overall detector control logic.
- The DetectorC module encapsulates all sensor-specific code, including switching the sensor on and off, and detecting loud sounds.
- The MicrophoneC module contains the low-level code for switching the mts300 microphone on and off.
- StatsC is a simple statistics module that provides a least-squares linear regression implementation (the code is a standard least-squares implementation and not shown in this book).
- The DetectorAppC configuration wires all the modules to each other and to the various system services.

The rest of this chapter describes the SynchronizerC, DetectorC, and MicrophoneC implementations in detail. These components are not portable, but the whole application is designed so that it can reasonably easily be ported to another mote or to another sensor (with different event detection logic).

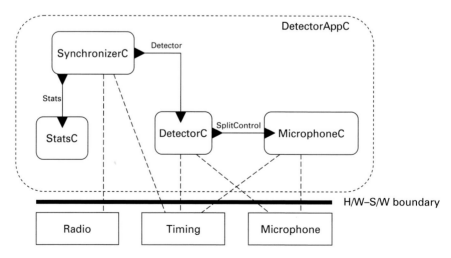

Figure 13.3 Detector mote components and hardware. Dashed lines show the hardware resources used by each component.

Several factors can cause SoundLocalizer to make an incorrect report of the mote closest to the sound. First, as we noted above, if the difference in travel times to two motes is less than the error in time synchronization, then those two motes will not reliably know which one is closer to the sound. Second, if the motes are sampling the microphone every 100 μs, then there will be an average delay of 50 μs in detecting a loud sound. Under the assumption that sampling doesn't start and stay in perfect synchrony on all motes, this delay increases SoundLocalizer's imprecision. Finally, if sound cannot travel in a straight line to the closest mote (e.g. there is an obstacle in the way), then the mote that hears the sound first may not be the closest mote.

Porting SoundLocalizer to a new mote that uses the same mts300 sensor board should only require significant changes in DetectorC: SynchronizerC and MicrophoneC use only hardware-independent interfaces that should just need rewiring for the new mote's implementation of the necessary hardware resources (timers, I/O pins, etc., as described below). SoundLocalizer can also be modified to use a different sensor, detecting a different event. For instance, the closest mote to an impact might be detectable using an accelerometer that measures vibrations. Modifying SoundLocalizer in this way will require a replacement for MicrophoneC to switch the new sensor on and off, and a replacement for DetectorC to implement the new event detection logic.

13.2 SynchronizerC

A detector mote goes through several execution stages: booting, time synchronization, preparing for event detection, detecting the event itself, cleaning up after detection, and voting. Execution then returns to the time synchronizing stage the next time the coordinator is switched on. The logic that controls the progression of a mote through these stages is in the SynchronizerC component and is portable to any platform that provides a microsecond-precision counter (used for time synchronization).

At boot time, a detector mote simply starts the radio and awaits messages from the coordinator (see above for the coordinator code and message layout). The handler for these coordination messages contains the heart of SoundLocalizer's time synchronization algorithm:

```
module SynchronizerC {
 uses interface Receive as RCoordination;
 uses interface Counter<TMicro, uint32_t>;
 uses interface Stats;
 ...
}
implementation {
 uint32_t t0;

 uint32_t now() {
    return call Counter.get() >> TIME_SHIFT;
 }
```

```
event message_t *RCoordination.receive(message_t *m, void *payload, uint8_t len){
  uint32_t arrivalTime = now(); // Note arrival time.

  if (len == sizeof(coordination_msg_t)) // Validate received message
    {
      coordination_msg_t *cmsg = payload;

      if (... first message from coordinator ...)
    {
      t0 = arrivalTime;
      call VotingTimer.stop();
      call Stats.reset();
    }

  call Stats.data(cmsg->count, arrivalTime - t0);

  // Prepare for event detection when its time is close
  if (cmsg->count >=
      cmsg->sample_at - (MICROPHONE_WARMUP / cmsg->interval) - 3)
    scheduleSampling(cmsg->sample_at);
    }
  return m;
}
...
}
```

Listing 13.1 SynchronizerC: time synchronization for SoundLocalizer

This code's first step is to note the message's arrival time by calling the get command of the Counter<TMicro, uint32_t> interface. Counter is a hardware-independent interface for counting time. Like other other time interfaces, Counter takes type arguments which indicate the precision (millisecond (TMilli), microsecond (TMicro), etc.), and width (16-bit (uint16_t), 32-bit (uint32_t), etc.) of its time measurements:

```
interface Counter<precision_tag, size_type> {
  async command size_type get();
  async command bool isOverflowPending();
  async command void clearOverflow();
  async event void overflow();
}
```

Listing 13.2 The Counter interface

The get command returns the current time, while the remaining commands and events allow a Counter user to gracefully handle counter overflow – a 32-bit microsecond counter overflows every 71 minutes, a 16-bit microsecond counter overflows 15 times a second, so handling overflow is a practical requirement for many applications. However, in the case of SoundLocalizer we can ignore the overflow issue. SynchronizerC divides

13.2 SynchronizerC

the returned microsecond time by 16 (TIME_SHIFT is 4) to reduce the range of values that the statistics package needs to handle.

TinyOS does not have a HIL component that provides a `Counter<TMicro, uint32_t>` interface. However, TinyOS does provide guidelines on the names of HAL components that offer these low-level time interfaces (see TEP102 [25]), so the following wiring for SynchronizerC should work on all platforms with a 32-bit microsecond counter:

```
components SynchronizerC, CounterMicro32C;
SynchronizerC.Counter -> CounterMicro32C;
```

The body of the coordination message handler uses the Stats interface (provided by SoundLocalizer's StatsC component) to collect statistics on the message receive times:

```
interface Stats {
  command void reset();
  command void data(uint32_t x, uint32_t y);

  command uint32_t count();
  command float estimateY(uint32_t x);
  command float estimateX(uint32_t y);
}
```

The reset command is called when the first message from the coordinator is received, and the data command is used to record the global time, local time sample pairs.

Finally, if the current global time (cmsg->count) is close to the global sampling time (cmsg->sample_at), then it's time to prepare for event detection by calling scheduleSampling. We need to leave enough time to warm up the sensor, and we add a cushion of two global time units in case some coordination messages get lost.

The scheduleSampling function converts the requested event detection time (sampleTime) into a local time by using the least-squares linear regression implemented by the estimateY command from the Stats interface:

```
void scheduleSampling(uint32_t sampleTime) {
  call Leds.led2Off();
  // We only agree to sample if we got enough coordination messages
  // to synchronize time properly
  if (call Stats.count() >= MIN_SAMPLES)
    {
      localSampleTime = call Stats.estimateY(sampleTime);
      // Stop the radio to avoid radio activity from interfering with
      // the precision of the event detection process
      call RadioControl.stop();
    }
}
```

Advanced application: SoundLocalizer

```
event void RadioControl.stopDone(error_t error) {
  // Once the radio is stopped, schedule event detection
  call Detector.start(t0 << TIME_SHIFT, localSampleTime << TIME_SHIFT);
}
```

SoundLocalizer switches the radio off in scheduleSampling, as radio activity could cause significant jitter in event detection. For instance, if an unrelated application uses the same radio channel as SoundLocalizer, then SoundLocalizer will receive and process (and hopefully ignore) this other application's messages. However, all this work will impact event detection, possibly causing some motes to detect a loud sound late, or even to miss a loud sound completely.

Actual event detection is implemented in the DetectorC component described below. SynchronizerC and DetectorC coordinate via the Detector interface:

```
interface Detector {
  command void start(uint32_t t0, uint32_t dt);
  async event void detected();
  event void done(error_t ok);
}
```

The start command requests event detection for time t0+dt – this way of specifying a time is consistent with the deadline-based approach used by TinyOS's timing interfaces (see Section 6.1.1). When DetectorC detects a loud sound, it immediately signals the detected event. To maximize responsiveness and minimize jitter, DetectorC's event detection happens in an interrupt handler, so the detected event is asynchronous. DetectorC signals the done event (in task context, with SUCCESS as argument) after it has finished cleaning up after event detection. Also, if an error occurs, DetectorC will signal done directly (with a FAIL argument).

SynchronizerC implements handlers for the detected and done events:

```
norace uint32_t localDetectTime;
uint32_t detectTime;

async event void Detector.detected() {
  localDetectTime = now();
}

event void Detector.done(error_t ok) {
  // If event detection was successful, convert local detection time to
  // a global time, and start broadcasting our detection reports
  if (ok == SUCCESS)
    {
      detectTime = 100000 * call Stats.estimateX(localDetectTime - t0);
      call VotingTimer.startPeriodic(VOTING_INTERVAL);
      call Leds.led2On(); // Assume we're closest
    }
  call RadioControl.start();
}
```

13.2 SynchronizerC

The only task of the asynchronous detected event is to store the precise local event detection time in the localDetectTime variable. Because detected runs in interrupt context, there might be a race on the write to the shared localDetectTime variable. However, the structure of SynchronizerC and DetectorC guarantees that the detected event is signaled before the done event which contains the only other access to localDetectTime. Thus, no data-race is possible on localDetectTime. To avoid warnings from the nesC compiler about possible races on localDetectTime we add a **norace** qualifier to localDetectTime's declaration.[1]

Once event detection is complete and the done event is signaled, SynchronizerC converts the local event detection time (localDetectTime) into a global event detection time (detectTime) using the Stats interface. Detection times are scaled by a factor of 100 000 to allow for precise comparisons of the detection time (detectTime is an unsigned integer). SynchronizerC also turns on the yellow LED to let the user know that a loud sound was detected, and starts the voting timer which will broadcast a mote's detection time. These broadcasts contain the mote's identity and global detection time:

```
typedef nx_struct detection_msg {
    nx_uint16_t id;
    nx_uint32_t time;
} detection_msg_t;
```

The broadcast and "voting" code is very straightforward (error-checking elided below for simplicity):

```
message_t msg;

event void VotingTimer.fired() {
  // Simply build and broadcast a detection message
  detection_msg_t *dmsg = call SDetection.getPayload(&msg, sizeof *dmsg);

  dmsg->id = TOS_NODE_ID;
  dmsg->time = detectTime;
  call SDetection.send(AM_BROADCAST_ADDR, &msg, sizeof *dmsg);
}

event message_t* RDetection.receive(message_t *m, void* payload, uint8_t len){
  detection_msg_t *dmsg = payload;

  // Voting logic: if received message indicates global detection time
  // earlier than ours, switch off our LED and stop broadcasting
  if (dmsg->time < detectTime)
    {
       call Leds.led2Off();
       call VotingTimer.stop();
    }
  return m;
}
```

---

[1] **norace** should be used with care, only when you are very sure that there is indeed no possible race.

13.3 DetectorC

DetectorC provides a Detector interface that detects loud sounds. Its internal logic is very simple. First, it powers up the microphone. Next, from the time specified in the Detector.start command it repeatedly samples the microphone's A/D channel until sound is above a threshold. Finally it powers off the microphone. The microphone is powered on and off by a SplitControl interface provided by the MicrophoneC component described below.

As we discuss below, DetectorC uses the micaz's HAL interface to the A/D converter. The A/D converter is a shared driver, with access controlled by a Resource interface (Section 11.2). Thus, when DetectorC's `start(t0, dt)` command is called, it must power up the microphone, request access to the A/D converter and schedule event detection for local time `t0+dt`:

```
module DetectorC {
  provides interface Detector;
  uses interface Alarm<TMicro, uint32_t>;
  uses interface Resource as AdcResource;
  uses interface SplitControl as Microphone;
  ...
}
implementation {
 bool granted, started; // Status of start request

 command void Detector.start(uint32_t t0, uint32_t dt) {
    atomic granted = started = FALSE;
    call Alarm.startAt(t0, dt);
    call Microphone.start();
    call AdcResource.request();
 }

 event void AdcResource.granted() {
    atomic granted = TRUE; // Note when ADC granted
 }

 event void Microphone.startDone(error_t error) {
    atomic started = error == SUCCESS; // Note if microphone started
 }

 async event void Alarm.fired() {
  // It's time to detect a loud sound. If we didn't get the ADC or
  // turn on the microphone in time, report a failed event detection.
    atomic
      if (granted && started)
        {
          call Leds.led1On();
          ... start detection ...
        }
      else
      post detectFailed();
 }
```

```
task void detectFailed() {
  call Leds.led0Toggle();
  signal Detector.done(FAIL);
 }
}
```

Listing 13.3 DetectorC: loud sound detection for SoundLocalizer

DetectorC uses an `Alarm<TMicro, uint32_t>` interface to schedule the event detection with microsecond precision. The Alarm interface is a low-level counterpart to Timer:

```
interface Alarm<precision_tag, size_type> {
  // basic interface
  async command void start(size_type dt);
  async command void stop();
  async event void fired();

  // extended interface
  async command bool isRunning();
  async command void startAt(size_type t0, size_type dt);
  async command size_type getNow();
  async command size_type getAlarm();
}
```

Listing 13.4 The Alarm interface

Like Counter, Alarm takes type arguments which indicate the precision (millisecond (TMilli), microsecond (TMicro), etc.) and width (16-bit (uint16_t), 32-bit (uint32_t), etc.) of its time units. Alarm is similar to the higher-level Timer interface (Section 6.1.1), except that it only offers one-shot timers (start, startAt), its commands can all be called from interrupt handlers, and its fired event runs in interrupt context (all the interface functions are **async**).

As with Counter, TinyOS does not have a HIL component that provides a `Alarm<TMicro, uint32_t>` interface, but TEP102 [25] also provides guidelines on the names of HAL components offering Alarm interfaces. Thus the following wiring for DetectorC should work on all platforms with a 32-bit microsecond alarm:

```
components new AlarmMicro32C() as DAlarm;
DetectorC.Alarm -> DAlarm;
```

Note that platforms may provide only a limited number of alarms, e.g. the micaz only supports three 32-bit microsecond alarms – each micaz Alarm needs its own hardware timer compare register, and the micaz's ATmega128 microcontroller only has three compare registers per hardware timer.

By simultaneously scheduling the alarm, microphone power on, and A/D converter request, DetectorC.start allows the execution of these three operations to overlap (assuming that they don't have overlapping resource requirements). The completion of the microphone power up and A/D converter request is tracked by the started and granted boolean variables respectively. Then Alarm.fired checks that these two booleans are true before initiating event detection. If they are not, DetectorC reports a failed detection back to SynchronizerC. The fired event runs in an interrupt context, thus there is a potential data race between the accesses to the started and granted variables in Alarm.fired, AdcResource.granted, and Microphone.startDone. This race risk is real (unlike that for localDetectTime in SynchronizerC), so DetectorC uses nesC's atomic statements to protect the reads and writes of granted and started.

To precisely detect when the loud sound occurs, DetectorC should sample the microphone at the highest possible rate. Furthermore, there should be the minimal amount of jitter between the time the microcontroller acquires the A/D sample and when DetectorC checks that this A/D value is above the "loud sound" threshold: adding, e.g., a random delay of 50–100 μs between sample acquisition and the threshold test adds an extra 1.7cm (50 μs · 340 m/s) of imprecision to SoundLocalizer. Both of these factors argue against using the high-level Read interface to sensors (Section 6.2.1): MicC (the mts300 sensorboard's microphone component) provides a Read that has a fairly high overhead, and which does a fair bit of processing after a sample has been acquired.

It would probably be possible to implement DetectorC using the ReadStream interface (Section 6.2.4) provided by the MicStreamC component. This interface offers high-rate low-jitter sampling, but is relatively complex to use and does not offer a direct indication of the exact time at which each sample was acquired. Instead, DetectorC uses Atm128AdcC, TinyOS's HAL A/D converter component for the ATmega128. Atm128AdcC is a shared driver (access is controlled by a Resource interface) and offers the following interface for low-level A/D access:

```
interface Atm128AdcSingle {
  async command bool getData(uint8_t channel, uint8_t refVoltage,
                bool leftJustify, uint8_t prescaler);
  async event void dataReady(uint16_t data, bool precise);
  async command bool cancel();
}
```

Listing 13.5 Atm128AdcSingle: low-level single-sample ATmega128 A/D converter interface

This interface allows A/D conversion requests to be made from interrupt handlers (getData is async), and signals sample acquisition (dataReady) directly from the ATmega128's A/D conversion interrupt handler. These two facts allow DetectorC to sample at a very high rate and with very low event detection jitter. Furthermore, the getData command exposes a prescaler parameter (not available via the Read or ReadStream interfaces) that adjusts the time it takes to perform an A/D conversion. Reducing the A/D conversion time reduces the precision of the A/D conversion, so most high-level users (including MicC and MicStreamC) leave the prescaler at the setting that guarantees full A/D conversion precision – at this setting a conversion takes 113 μs on a micaz mote. However, detecting when a value goes above a threshold does not require the full A/D precision (only the high-order bits are really relevant), so

DetectorC sets the prescaler to a value (ATM128_ADC_PRESCALE_16) that reduces A/D conversion time to 28 μs. This potentially improves SoundLocalizer's precision by 2.9 cm (85 μs · 340 m/s).

The resulting "loud sound" detection code is simple, though very specific to the micaz platform and the mts300 microphone:

```
module DetectorC {
  uses interface Atm128AdcSingle;
  ...
}
implementation {
  // The threshold for a loud sound (determined experimentally)
  enum { THRESHOLD = 768 };

  void detect() {
    call Atm128AdcSingle.getData(... microphone A/D channel ...,
                     ATM128_ADC_VREF_OFF, FALSE,
                     ATM128_ADC_PRESCALE_16);
  }


  async event void Atm128AdcSingle.dataReady(uint16_t data, bool precise){
    /* If we're the current ADC owner: check ADC completion events to
       see if the microphone is above the threshold */
    atomic
      if (granted)
        if (precise && data > THRESHOLD)
          {
             signal Detector.detected();
             ... post a task to release A/D converter, power off microphone
                 and signal done ...
          }
        else
          detect();
  }
  ...
}
```

The dataReady event handler checks that DetectorC currently owns the A/D converter. If so, A/D results above the threshold indicate that a loud sound was detected. If not, another A/D sample is requested.

13.4 MicrophoneC

MicrophoneC is a fairly typical example of low-level code to control a sensor. The mts300 microphone hardware is shown schematically in Figure 13.4. The microphone is powered on and off via a digital I/O pin (MicPower) from the microcontroller. The

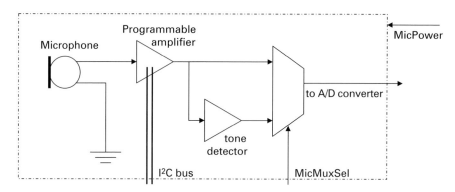

Figure 13.4 MTS300 microphone hardware.

raw microphone output goes through an amplifier, whose gain can be adjusted via an I²C bus. Finally, a second digital I/O pin (MicMuxSel) selects between raw microphone output and a 4 kHz tone detector.[2]

Powering on the microphone for use as a "loud sound" detector thus requires physically powering on the microphone, selecting raw microphone output and picking an appropriate amplifier gain. Finally, once these steps are complete, the microphone should be left to warm up for 1.2 s (MICROPHONE_WARMUP). These four steps are performed by the following code from MicrophoneC (error checking elided):

```
module MicrophoneC {
  provides interface SplitControl;
  uses interface GeneralIO as MicPower;
  uses interface GeneralIO as MicMuxSel;
  uses interface I2CPacket<TI2CBasicAddr>;
  uses interface Resource as I2CResource;
}
implementation {
  enum {
    MIC_POT_ADDR = 0x5A,   // Amplifier I2C address
    MIC_POT_SUBADDR = 0,
    MIC_GAIN = 64          // Desired gain
  };

  command error_t SplitControl.start() {
    // Power up the microphone
    call MicPower.makeOutput();
    call MicPower.set();
    // Select raw microphone output
    call MicMuxSel.makeOutput();
    call MicMuxSel.set();
    // Request the I2C bus to adjust gain
```

---

[2] The mts300 sensor board was originally designed in part for use in acoustic-based localization, and includes a 4 kHz tone generator.

```
  call I2CResource.request();

  return SUCCESS;
}

event void I2CResource.granted() {
  static uint8_t gainPacket[2] = { MIC_POT_SUBADDR, MIC_GAIN };

  // Send gain-control packet over I2C bus
  call I2CPacket.write(I2C_START | I2C_STOP, MIC_POT_ADDR,
                       sizeof gainPacket, gainPacket);
}

async event void I2CPacket.writeDone(error_t error, uint16_t addr,
                                     uint8_t length, uint8_t* data) {
  // Release I2C bus and wait for microphone to warm up (report failure
  // in case of error)
  call I2CResource.release();
  post gainOk(); // We're in async code, post a task for the warm up timer
}

task void gainOk() {
  call Timer.startOneShot(MICROPHONE_WARMUP);
}

event void Timer.fired() {
  // Microphone warmed up. Signal completion of startup.
  signal SplitControl.startDone(SUCCESS);
}
}
```

This microphone power up code is of course specific to the mts300 sensor. However, it is actually quite portable to other platforms using the same sensor as it accesses the required low-level hardware features (digital I/O pins and the I^2C bus) via the hardware-independent GeneralIO and I2CPacket interfaces. The GeneralIO interface offers commands to configure, read and write a typical microcontroller digital I/O pin:

```
interface GeneralIO {
  async command void set();
  async command void clr();
  async command void toggle();
  async command bool get();
  async command void makeInput();
  async command bool isInput();
  async command void makeOutput();
  async command bool isOutput();
}
```

Listing 13.6 The GeneralIO digital I/O pin interface

In TinyOS, an I²C bus is normally accessed via a shared driver, so I²C bus users first request access via a Resource interface (named I2CResource in MicrophoneC). The I2CPacket interface provides commands for I²C bus masters to send read and write packets (I2CPacket takes an addr_size type argument to distinguish I²C implementations with 7 vs 15-bit bus addresses):

```
interface I2CPacket<addr_size> {
 async command error_t read(i2c_flags_t flags, uint16_t addr, uint8_t length, uint8_t* data);
 async event void readDone(error_t error, uint16_t addr, uint8_t length, uint8_t* data);

 async command error_t write(i2c_flags_t flags, uint16_t addr, uint8_t length, uint8_t* data);
 async event void writeDone(error_t error, uint16_t addr, uint8_t length, uint8_t* data);

}
```

Listing 13.7 The I2CPacket interface for bus masters

While MicrophoneC is quite portable, the wiring that connects it to the specific hardware resources used to access the microphone is necessarily platform-specific:

```
components MicaBusC, new Atm128I2CMasterC() as I2CPot;
MicrophoneC.MicPower    -> MicaBusC.PW3;
MicrophoneC.MicMuxSel   -> MicaBusC.PW6;
MicrophoneC.I2CResource -> I2CPot;
MicrophoneC.I2CPacket   -> I2CPot;
```

Atm128I2CMasterC is the component that provides access to the ATmega128's I²C bus. MicaBusC abstracts the differences between the various mica-family motes that share the same 51-pin sensor board connector, making it easier to write sensor board implementations portable to all mica-family motes.

Powering off the microphone is much simpler. The MicPower I/O pin just needs to be cleared and made into an input:

```
command error_t SplitControl.stop() {
  //Power off microphone
  call MicPower.clr();
  call MicPower.makeInput();

  //And let our caller know we're done-post a task as one should not
  //signal events directly from commands
  post stopDone();
  return SUCCESS;
}

task void stopDone() {
  signal SplitControl.stopDone(SUCCESS);
}
```

13.5 Wrap-up

That's it! You've seen how to build applications, from the very simplest version of anti-theft (Chapter 6.1) to moderately complex systems like SoundLocalizer. You've also seen how nesC's component and execution model, and TinyOS's component libraries help you structure applications and build reusable abstractions (libraries, services, device drivers). Remember to check `www.tinyos.net` for the latest developments and documentation on TinyOS and nesC.

Now it's up to you to go and write some useful and interesting sensor network applications!

Part IV

Appendix and references

Appendix TinyOS APIs

This appendix gives summaries of major TinyOS interfaces and the components that provide them. It only covers hardware-independent HIL interfaces and abstractions that are common across multiple hardware platforms. In almost all cases, TinyOS Enhancement Proposals (TEPs) describe the abstractions in greater depth, and we reference them as appropriate. Application-level abstractions with split-phase interfaces are typically generics, which a user must instantiate. For sake of brevity, this list shows the signatures of relevant components: you can read the interfaces in your TinyOS distribution.

A.1 Booting

The process of booting a mote is encapsulated in the MainC component, that signals the Boot.booted event once all system services are initialized. Components that require boot-time initialization should wire themselves to MainC's SoftwareInit: MainC's SoftwareInit.init command is called after hardware initialization but before booted is signaled.

```
configuration MainC {
  provides interface Boot;
  uses interface Init as SoftwareInit;
}
```

We covered booting in Section 6.1. The TinyOS boot process is described in TEP 107: TinyOS 2.x Boot Sequence [14].

A.2 Communication

TinyOS provides four basic communication abstractions: active messages (AM, single-hop, unreliable communication), collection (multi-hop, unreliable delivery to a data sink) and dissemination (multi-hop, reliable delivery to every node in the network), and serial communication (unreliable delivery over a serial line). We covered communication in Sections 6.3, 6.4, and 7.1.1.

The message_t type is used to declare packet buffers for TinyOS networking protocols, as explained in TEP 111: message_t [15].

A.2.1 Single-hop

TinyOS has four standard single-hop communication components, all of which take an AM type as a parameter, an 8-bit number that identifies different packet types. Active message addresses are 16 bits. TinyOS defines the macro TOS_BCAST_ADDR for the AM broadcast address, and a node can find its local address through the AMPacket interface.

AMSenderC sends packets. AMReceiverC signals reception for packets addressed to the node, including the broadcast address; AMSnooperC signals reception for packets not addressed for the node; AMSnoopingReceiverC signals reception for all packets, regardless of addressing. The AM interfaces automatically discard packets which do not pass a link-layer CRC check. The PacketAcknowledgements interface allows transmitters to request for a link-layer acknowledgement and check if one was received after transmission: PacketAcknowledgements.wasAcked can be called in Send.sendDone.

```
generic configuration AMSenderC(am_id_t AMId) {
  provides {
    interface AMSend;
    interface Packet;
    interface AMPacket;
    interface PacketAcknowledgements as Acks;
  }
}

generic configuration AMReceiverC(am_id_t amId) {
  provides {
    interface Receive;
    interface Packet;
    interface AMPacket;
  }
}

generic configuration AMSnooperC(am_id_t AMId) {
  provides {
    interface Receive;
    interface Packet;
    interface AMPacket;
  }
}

generic configuration AMSnoopingReceiverC(am_id_t AMId) {
  provides {
    interface Receive;
    interface Packet;
```

```
    interface AMPacket;
  }
}
```

By default, the radio stack is powered off. You must turn it on in order to be able to send and receive packets. The simplest way to do so is to put it in full-power (always-on) mode via the SplitControl interface. The low-power section below (page 251) shows how you can turn it on into a low-power state, where nodes periodically wake up to listen for packets. For full-power control, wire and call SplitControl.start on ActiveMessageC.

```
configuration ActiveMessageC {
  provides {
    interface SplitControl;
    // interfaces elided here
  }
}
```

For more details on these components and their interfaces, refer to TEP 116: Packet Protocols [16]. The AM components are typically found in `tos/system`.

The above components are all for radio communication: there are also versions of these components for motes that are connected via a serial port: SerialActiveMessageC, SerialAMSenderC, and SerialAMReceiverC. These components are described in TEP 113: Serial Communication [7]. Note that the serial stack does not perform address filtering, so there is no snooping.

A.2.2 Multi-hop collection

Collection protocols build routing trees to nodes that advertise themselves as data sinks. Typically, sink nodes are connected to a PC via a serial cable or other medium, although sometimes they log all data they receive to non-volatile storage. While in good conditions collection protocols can have 99.9% or better reliability, they do not promise reliability: a sender does not receive feedback on whether its packets have arrived.

Collection implementations can be found in `tos/lib/net`. For example, the collection tree protocol (CTP) implementation can be found in `tos/lib/net/ctp` while the MultihopLQI implementation can be found in `tos/lib/net/lqi`.

To send packets, a program instantiates a CollectionSenderC. To configure a node as a base station, it should wire to CollectionControlC to call RootControl.setRoot as well as instantiate relevant CollectionReceiverC services. Collection clients take a collection identifier, similar to an AM type.

```
generic configuration CollectionSenderC(collection_id_t collectid) {
  provides {
    interface Send;
    interface Packet;
  }
}

generic configuration CollectionReceiverC(collection_id_t collectid) {
  provides {
```

```
    interface Receive;
  }
}

configuration CollectionControlC {
  provides {
    interface StdControl;
    interface Packet;
    interface CollectionPacket;
    interface RootControl;
  }
}
```

Note that you must start collection by calling CollectionControlC.StdControl. Further details on these components and collection can be found in TEP 119: Collection [2].

A.2.3 Multi-hop dissemination

Collection pulls data out of a network: dissemination pushes it into a network. Dissemination protocols are reliable and deliver data to every node in a network: new nodes that join the network will receive updates. Achieving this reliability requires that a node statically allocate RAM for the data, so it can forward it at any time. Dissemination is typically used for controlling and configuring a network: a dissemination value can be a configuration constant, or a concise command. Dissemination only works for small values that can fit in a single packet (e.g. 20 bytes). Each dissemination value has a unique key, which a user must specify.

Dissemination implementations can be found in `tos/lib/net`. Drip [27] in `tos/lib/net/drip` is the simpler of the two: it requires less RAM and code space but is less efficient. DIP, in `tos/lib/net/dip` is more efficient but more complex. The DisseminationValue interface notifies a node of updates, while the DisseminationUpdate interface lets a node create an update. Configuring a network from a PC typically involves sending a packet to a base station over a serial port that tells the node to update a value.

Both DIP and Drip use the Trickle algorithm [17]. Trickle values efficiency over speed: it can take a few seconds for a value to propagate one hop, and it might take up to a minute to disseminate to every node in a very large, many-hop network. It works slowly to avoid flooding the network with lots of packets and causing many collisions.

```
generic configuration DisseminatorC(typedef t, uint16_t key) {
  provides interface DisseminationValue<t>;
  provides interface DisseminationUpdate<t>;
}

configuration DisseminationC {
  provides interface StdControl;
}
```

Note that you must start dissemination by calling DisseminationC.StdControl. Further details on dissemination can be found in TEP 118: Dissemination [18].

A.2.4 Binary reprogramming

The fourth common communication abstraction is binary reprogramming. This is not really a abstraction, in that an application doesn't actively send packets using it. Instead, it is a service that automatically runs. It enables an administrator to install new binary images in a TinyOS network. These binaries are stored on flash, and can be installed. Note that if a new image doesn't have binary reprogramming installed, then you can't uninstall it!

Deluge, found in `tos/lib/net/Deluge`, is the standard binary dissemination implementation [9]. It builds on top of basic dissemination to install large (many kB) data items. Rather than have TinyOS APIs for controlling its behavior, Deluge comes with command line tools that inject the necessary packets into a network. Please refer to the Deluge manual in the TinyOS documentation for more details.

A.3 Time

TinyOS has two timer interfaces: Timer, which is synchronous and operates in task context, and Alarm, which is asynchronous and involves directly handling a hardware interrupt. For the most part, applications use Timer. Some low-level systems (such as MAC protocols) or applications that require precise timing use Alarm. Additionally, nodes can access their local time with the LocalTime interface. A node's local time starts when it boots, so different nodes typically do not have the same time.

The basic Timer abstraction is TimerMilliC:

```
generic configuration TimerMilliC() {
  provides interface Timer<TMilli>;
}
```

Local time can be obtained from LocalTimeMilliC:

```
configuration LocalTimeMilliC {
  provides interface LocalTime<TMilli>;
}
```

Because Alarm is typically a HAL, rather than HIL, abstraction (Chapter 12), there is no standard component that provides Alarm. We covered timing in Section 6.1 and Chapter 13. Further details on timers and time can be found in TEP 102: Timers [25].

A.4 Sensing

Sensors in TinyOS follow a naming convention described in TEP 109: Sensors and Sensor Boards [5]. Typically, they are named after the actual sensor chip. Therefore, there are almost no "standard" sensor component names, as the set of a sensors a node might have depends on its hardware. However, TEP 109 also describes what interfaces sensors should provide, so that it's typically easy to write code independently of the exact sensor.

Sensors normally provide one or more of the common sensing interfaces described in TEP 114: SIDs: Source and Sink Independent Drivers [28].

TinyOS provides two "fake" sensors, which are completely software. The first, ConstantSensorC, takes a constant as a parameter and always returns that value. The second, SineSensorC, returns a value from a sine function whose input increments on each sample. Platforms also have a DemoSensorC, which either instantiates one of these software sensors or is a wrapper around a hardware-specific sensor.

```
generic module ConstantSensorC(typedef width_t @integer(), uint32_t val) {
  provides interface Read<width_t>;
}

generic module SineSensorC() {
  provides interface Read<uint16_t>;
}

generic configuration DemoSensorC() {
  provides interface Read<uint16_t>;
}
```

We covered sensors in Section 6.2 and Chapter 13.

A.5 Storage

TinyOS provides three basic storage abstractions. Config storage is for small, random-access variables, such as configuration constants. Log storage is for append-only writing and random-access, sequential reading. Block storage is for large, random-access data items. Section 6.5 has more details on their use and tradeoffs.

TinyOS has scripts that generate a layout of named storage volumes from an XML specification. A layout is essentially the offset where each volume starts and its length. Storage clients take a volume ID as a parameter, which allows it to access these generated constants. For more details, refer to TEP 103: Permanent Data Storage (Flash) [4].

```
generic configuration BlockStorageC( volume_id_t volume_id ) {
  provides interface BlockRead;
  provides interface BlockWrite;
  provides interface StorageMap;
}

generic configuration LogStorageC( volume_id_t volume_id, bool circular ) {
  provides interface LogRead;
  provides interface LogWrite;
}
```

```
generic configuration ConfigStorageC( volume_id_t volume_id ) {
  provides interface Mount;
  provides interface ConfigStorage;
}
```

Because flash storage implementations are chip-specific, you usually find them in a `tos/chips` directory. For example, the mica family of nodes has an AT45DB-family flash chip: when you use one of the above abstractions on a mica, you use the components in `tos/chips/at45db`.

A.6 Data structures

TinyOS has component implementations of a few commonly used data structures. All of these can be found in `tos/system`.

A.6.1 BitVectorC

BitVectorC provides the abstraction of a bit vector. It takes a single parameter, the width of the vector. The BitVector interface has commands for getting, setting, clearing, and toggling individual bits or all of the bits at once. Because of nesC's heavy inlining, using BitVectorC is preferable to writing your own bit vector macros within a component. BitVectorC allocates $\lceil \frac{N}{8} \rceil$ bytes for an N-bit wide vector.

```
generic module BitVectorC(uint16_t max_bits) {
  provides interface BitVector;
}
```

A.6.2 QueueC

QueueC provides the abstraction of a queue of items with a fixed maximum size. It takes two parameters: the type of the items it stores and the maximum size of the queue. The Queue interface has commands for enqueuing items on the end of the queue, dequeuing the head of the queue, and commands for checking the queue size. It also allows random-access lookup: you can scan the queue.

```
generic module QueueC(typedef queue_t, uint8_t QUEUE_SIZE) {
  provides interface Queue<queue_t>;
}
```

QueueC is used heavily in networking protocols. A routing protocol, for example, often creates a QueueC of pointers to message buffers (`message_t*`) for its forwarding queue, as well as a PoolC (see below) to allocate a number of buffers so it can receive packets to forward.

A.6.3 BigQueueC

The uint8_t parameter to QueueC limits its maximum size to 255. For most uses, this is sufficient and so wasting extra bytes on its internal fields to support a larger size is not worth it. However, sometimes components need a larger queue. The printf library, for example (page 250), has a queue of bytes to send, which is usually longer than 255. TinyOS therefore also has BigQueueC, which is essentially identical to QueueC except that it has a 16-bit size parameter and provides the interface BigQueue:

```
generic module BigQueueC(typedef queue_t, uint16_t QUEUE_SIZE) {
  provides interface BigQueue<queue_t> as Queue;
}
```

A.6.4 PoolC

PoolC is the closest thing TinyOS has to a dynamic memory allocator. It takes two parameters: the type of object to allocate, and how many. Components can then dynamically allocate and free these objects to the pool. But as the maximum pool size is set at compile-time, a memory leak will cause the pool to empty, rather than cause the heap and stack to collide.

Because it does the allocation, you specify a type to PoolC, but its commands use pointers to that type. For example, if you allocate a pool of message_t buffers, then calls to Pool pass message_t*.

```
generic configuration PoolC(typedef pool_t, uint8_t POOL_SIZE) {
  provides interface Pool<pool_t>;
}
```

You can swap data items between pools. For example, if you have two separate message_t pools P_1 and P_2, it is OK to allocate M_1 from P_1 and M_2 from P_2, yet free M_1 into P_2 and M_2 into P_1. This behavior is critically important due to the buffer-swapping semantics of Receive. If a component receives a packet it wants to forward, it allocates a buffer from its pool and returns this new buffer to the link layer. The next packet – whose buffer came from the pool – might go to another component, which has its own pool. So the pseudocode for forwarding a packet looks something like this:

```
receive(m):
  if (!forward(m)):
    return m
  setHeaders(m)
  queue.put(m)
  m2 = pool.get()
  return m2
```

A.6.5 StateC

StateC provides an abstraction of a state machine. This is useful when multiple components need to share a global state (such as whether the subsystem is on or off).

```
generic configuration StateC() {
  provides interface State;
}
```

A.7 Utilities

TinyOS provides components for several commonly used functions and abstractions.

A.7.1 Random numbers

RandomC provide the interface Random, which components can use to generate random numbers. RandomC also provides the interfaces Init and ParameterInit, to enable a component to re-initialize RandomC's random seed. By default, RandomC's seed is initialized to the node ID+1.

TinyOS includes two random number generators: RandomMlcgC (a multiplicative linear congruential generator) and RandomLfsrC (a linear feed shift register generator). RandomLfsrC is faster, but RandomMlcgC produces better random numbers. By default, RandomC refers to RandomMlcgC.

```
configuration RandomC {
  provides interface Init;
  provides interface ParameterInit<uint16_t> as SeedInit;
  provides interface Random as Random;
}

configuration RandomMlcgC {
  provides interface Init;
  provides interface ParameterInit<uint16_t> as SeedInit;
  provides interface Random as Random;
}

configuration RandomLfsrC {
  provides interface Init;
  provides interface ParameterInit<uint16_t> as SeedInit;
  provides interface Random as Random;
}
```

A.7.2 Leds

LedsC provides an abstraction of three LEDs. While some platforms have more or fewer than three, the Leds interface has three for historical reasons. Also, breaking up the LEDs into three instances of the same interface would be a lot of extra wiring. In addition to LedsC, there is also a NoLedsC, which can be dropped in as a null replacement: calls to NoLedsC do nothing.

```
configuration LedsC {
  provides interface Leds;
}

configuration NoLedsC {
  provides interface Leds;
}
```

A.7.3 Cyclic redundancy checks

Cyclic redundancy checks (CRCs) are a simple way to check whether a piece of data has been corrupted. After the data, you append a CRC. Someone reading the data can recompute the CRC over the data and check that it matches the appended CRC. If the two do not match, there is a bit error either in the data or the CRC itself. Since the CRC is usually much shorter than the data, the assumption is the data has been corrupted. CRCs are heavily used in networking, to check the validity of a packet. Of course, since CRCs are shorter than the data itself, it's possible, but unlikely, for a corrupted packet to pass a CRC check.

CRC values are distinct from cryptographic hashes. CRCs are intended to detect bursts of bit errors. Typically, an n-bit CRC can always detect a single error burst that is shorter than n bits. In contrast, cryptographically strong hashes have entropy properties that make detecting (or failing to detect) any kind of error uniformly likely.

```
module CrcC {
  provides interface Crc;
}
```

The module CrcC can be found in `tos/system`.

A.7.4 Printf

Sometimes, when debugging, it can very useful to have a mote send simple text messages. TinyOS has a printf library – like the C standard library function – for this purpose. You can use printf in your components, and the printf library will send appropriate packets over the serial port. You must start the printf library via PrintfC's SplitControl.start.

```
configuration PrintfC {
  provides {
      interface SplitControl as PrintfControl;
      interface PrintfFlush;
  }
}
```

For more details on using the printf library, refer to the tutorial on the TinyOS website.

A.8 Low power

For the most part, TinyOS will automatically put systems and hardware into the lowest power state possible. For peripherals, this typically works through power locks (Section 11.2). For example, when an application writes or reads from a flash chip, TinyOS will automatically power it on, perform the operation, then power it down when done.

Communication is the major exception to this behavior. Because communication subsystems, such as the radio and serial port, need to be able to receive as well as transmit, TinyOS needs an application to explicitly tell it when it can safely turn them on and off via SplitControl interfaces.

In the case of the radio, however, there are many techniques one can use to save power. For example, rather than always keep the radio on, TinyOS can keep it off most of the time, periodically turning it on just long enough to hear if there's a packet. A transmitter sends a packet multiple times, until the receiver wakes up, hears it, and acknowledges it, or a timeout occurs.

A complete discussion of how to set these intervals is beyond the scope of the book. The important interface is LowPowerListening, which some (but not all) radios provide. LowPowerListening allows an application to set a radio's check interval and also the check interval it expects of a receiver. TinyOS has a tutorial for how to use this interface when writing a low-power application, and Section 12.2.3 has a brief example of its use.

References

[1] J. Elson, L. Girod, and D. Estrin. Fine-grained network time synchronization using reference broadcasts. In *OSDI '02: Fifth Symposium on Operating Systems Design and Implementation*, Boston, MA, USA, Dec 2002.

[2] R. Fonseca, O. Gnawali, K. Jamieson, and P. Levis. TEP 119: Collection. www.tinyos.net/tinyos-2.x/doc/.

[3] E. Gamma, R. Helm, R. Johnson, and J. Vlissides. *Design Patterns: Elements of Reusable Object-Oriented Software*. Addison-Wesley, 1994.

[4] D. Gay and J. Hui. TEP 103: Permanent Data Storage (Flash). www.tinyos.net/tinyos-2.x/doc/.

[5] D. Gay, P. Levis, W. Hong, J. Polastre, and G. Tolle. TEP 109: Sensors and Sensor Boards. www.tinyos.net/tinyos-2.x/doc/.

[6] J. Gosling, B. Joy, G. Steele, and G. Bracha. *The Java Language Specification (Third Edition)*. Prentice-Hall, 2005.

[7] B. Greenstein and P. Levis. TEP 113: Serial Communication. www.tinyos.net/tinyos-2.x/doc/.

[8] V. Handziski, J. Polastre, J.-H. Hauer, C. S. A. Wolisz, D. Culler, and D. Gay. TEP 2: The Hardware Abstraction Architecture. www.tinyos.net/tinyos-2.x/doc/.

[9] J. W. Hui and D. Culler. The dynamic behavior of a data dissemination protocol for network programming at scale. In *SenSys '04: Proceedings of the Second Conference on Embedded networked International sensor Systems*, pages 81–94, New York, NY, USA, 2004. ACM Press.

[10] S. Kim, S. Pakzad, D. Culler, J. Demmel, G. Fenves, S. Glaser, and M. Turon. Health monitoring of civil infrastructures using wireless sensor networks. In *IPSN '07: Proceedings of the Sixth International Conference on Information Processing in Sensor Networks*, 2007.

[11] K. Klues, V. Handziski, J.-H. Hauer, and P. Levis. TEP 115: Power Management of Non-Virtualised Devices. www.tinyos.net/tinyos-2.x/doc/.

[12] K. Klues, V. Handziski, C. Lu, A. Wolisz, D. Culler, D. Gay, and P. Levis. Integrating concurrency control and energy management in device drivers. In *SOSP '07: Proceedings of the Twenty-first ACM SIGOPS Symposium on Operating Systems Principles*, pages 251–264, New York, NY, USA, 2007. ACM Press.

[13] K. Klues, P. Levis, D. Gay, D. Culler, and V. Handziski. TEP 108: Resource Arbitration. www.tinyos.net/tinyos-2.x/doc/.

[14] P. Levis. TEP 107: TinyOS 2.x Boot Sequence. www.tinyos.net/tinyos-2.x/doc/.

[15] P. Levis. TEP 111: message_t. www.tinyos.net/tinyos-2.x/doc/.

[16] P. Levis. TEP 116: Packet Protocols. www.tinyos.net/tinyos-2.x/doc/.

[17] P. Levis, N. Patel, D. Culler, and S. Shenker. Trickle: A self-regulating algorithm for code maintenance and propagation in wireless sensor networks. In *NSDI '04: First USENIX/ACM Symposium on Network Systems Design and Implementation*, 2004.

[18] P. Levis and G. Tolle. TEP 118: Dissemination. www.tinyos.net/tinyos-2.x/doc/.

[19] J. Liu, N. Priyantha, F. Zhao, C.-J. M. Liang, and Q. Wang. Towards discovering data center genome using sensor nets. In *EmNets '08: Proceedings of the Fifth Workshop on Embedded Networked Sensors*, 2008.

[20] S. Madden, M. J. Franklin, J. M. Hellerstein, and W. Hong. Tinydb: An acquisitional query processing system for sensor networks. *Transactions on Database Systems (TODS)*, **30**(1), 2005, 122–173.

[21] A. Mainwaring, J. Polastre, R. Szewczyk, D. Culler, and J. Anderson. Wireless sensor networks for habitat monitoring. In *Proceedings of the ACM International Workshop on Wireless Sensor Networks and Applications*, 2002.

[22] M. Maroti, B. Kusy, G. Simon, and A. Ledeczi. The flooding time synchronization protocol. In *SenSys '04: Proceedings of the Second International Conference on Embedded Networked Sensor Systems*, pages 39–49, New York, NY, USA, 2004. ACM Press.

[23] J. Polastre, J. Hill, and D. Culler. Versatile low power media access for wireless sensor networks. In *SenSys '04: Proceedings of the Second International Conference on Embedded Networked Sensor Systems*, pages 39–49, New York, NY, USA, 2004. ACM Press.

[24] D. Rand. PPP Reliable Transmission. Internet Network Working Group RFC1663, July 1994.

[25] C. Sharp, M. Turon, and D. Gay. TEP 102: Timers. www.tinyos.net/tinyos-2.x/doc/.

[26] R. Szewczyk, J. Polastre, A. Mainwaring, and D. Culler. An analysis of a large scale habitat monitoring application. In *SenSys '04: Proceedings of the Second International Conference on Embedded Networked Sensor Systems*, New York, NY, USA, 2004. ACM Press.

[27] G. Tolle and D. Culler. Design of an application-cooperative management system for wireless sensor networks. In *EWSN '05: Proceedings of Second European Workshop on Wireless Sensor Networks*, 2005.

[28] G. Tolle, P. Levis, and D. Gay. TEP 114: SIDs: Source and Sink Independent Drivers. www.tinyos.net/tinyos-2.x/doc/.

[29] T. von Eicken, D. E. Culler, S. C. Goldstein, and K. E. Schauser. Active messages: A mechanism for integrated communication and computation. In *ISCA '92: Proceedings of the International Symposium on Computer Architecture*, pages 256–266, 1992.

[30] G. Werner-Allen, K. Lorincz, J. Johnson, J. Leess, and M. Welsh. Fidelity and yield in a volcano monitoring sensor network. In *OSDI '06: Seventh USENIX Symposium on Operating Systems Design and Implementation*, 2006.

Index

<− wiring operator, 31, **51**
−> wiring operator, 31, **51**
= wiring operator, 31, **52**
@ (declaring, using attributes), 142
@C attribute, 143
@atleastonce attribute, 143
@atmostonce attribute, 143, 212
@atomic_hwevent attribute, 143
@combine attribute, 66, 143
@exactlyonce attribute, 143
@hwevent attribute, 143
@integer attribute, 132, 143
@number attribute, 133, 143
@spontaneous attribute, 143
#define, 41, 45
   pitfalls, 46
#include, usage of, 45
__attribute__ (deprecated), 144
802.15.4, 5

abstract data type, 133
   implemented using generic modules, 133
   implemented using reference parameters, 134
   predefined in TinyOS , 247
active message type, 89, 94, 98, 113, 242
Active messages (AM)
   address and TOS_NODE_ID, difference, **90**
active messages (AM), 89, 113, 241
ActiveMessageC component, 137
Alarm interface, 231
AM type, *see* active message type
AMSend interface, 90
AntiTheft (application), 79
application source code, 9
as
   naming components, 53
   naming interfaces, 23
asynchronous (async)
   code, **71**, 192
   commands and events, **71**, 192
   consistency issues, 134
   use to minimize jitter, 228
Atm128AdcSingle interface, 232
atomic statement, **195**
   execution time, 196
   implemented by disabling interrupts, 195
   limitations, 200
   use in SoundLocalizer, 231
attribute, **142**
auto-wiring (for initialization), **60**, 139

backing array (for packets), **114**, 115, 123, 124
base station, **95**, 98
bidirectional interface, **28**
big-endian, 43
binary reprogramming, 245
BlockRead interface, 108
BlockWrite interface, 106
Boot interface, 81

C libraries, using, 47
C++ templates vs generic components, 130, 132
callback, 11, 15, 167
CC1000 (radio chip), **49**
CC2420 (radio chip), **5**
collection (network protocol), 95, 96, 98, 241
combine function, **66**
   associative and commutative, 67
   warning (when missing), 67
command, 11, **25**
   unique, *see* unique (compile-time function)
   uniqueCount, *see* uniqueCount (compile-time function)
component, 6, 11, 14, **21**
   implementation, 21, **29**
   initialization, 59
   layering, 60
   libraries, 162
   naming using as, 53
   signature, **21**
   singleton, **33**, 70
   switching between similar, 94
concurrency model, **71**, 192
ConfigStorage interface, 103
configuration, 12, 31, **50**
   enum and typedef in body, 150
   exporting interfaces, **52**
   generic, 68, 130, **150**

configuration (cont.)
    use in Facade pattern, 183
    use in Placeholder pattern, 180
constants, 41
Counter interface, 226
CRC (cyclic redundancy check), 250

data race, **193**
    automatic detection, 196, 229, 231
    avoiding, 196
deadline-based timing, 81
Deluge (binary reprogramming), 245
design patterns, 166
    Adapter, **189**
    behavioral, 166, 186, 189
    Decorator, **186**
    Dispatcher, **166**
    Facade, **183**
    Keymap, **177**
    Keyspace, **174**
    namespace, 174, 177
    Placeholder, **180**
    Service Instance, **170**
    structural, 170, 180, 183
device driver, 210
    access control, 211
    and the Resource interface, 214
    dedicated, 211
    latency, 213
    power management, 215
    shared, 211
    virtualized, 211
dissemination (network protocol), 95, 97, 98, 241
DisseminationUpdate interface, 99
DisseminationValue interface, 97
dynamic memory allocation, 37, 248

energy management, *see* power management
enum
    in configuration, 150
    use and abuse, 41
    use in Global Keyspace pattern, 174
event, 11, **25**
execution model, **71**
exporting (an interface), **52**

fan-in, **64**
fan-out, **64**
FlashSampler (application), 105

GeneralIO interface, 235
generic component, **33**, 68, 129, 150
    implemented by code copying, 130
    type arguments, 132
    use in Adapter pattern, 189

use in Decorator pattern, 187
use in Dispatcher pattern, 169
use in Service Instance pattern, 171
vs C++ template, 130, 132
generic interface, **27**
global declarations, 45

hardware abstraction architecture (HAA), **206**, 221
    storage, 208
    timers, 209
hardware adaptation layer (HAL), **206**
hardware interface layer (HIL), **206**, 241
hardware presentation layer (HIL), **207**
header files, 45
HIL, *see* hardware abstraction architecture
HIL, *see* hardware adaptation layer (HAL)

I^2C , 235
I2CPacket interface, 236
IEEE 802.15.4, 5
interface, 11, 14, **24**, 156
    bidirectional, **28**
    default command, event handlers, 142
    generic, **27**
    naming using as, 23
    parameterized, *see* parameterized interface
    provider, 11, **22**
    split-phase, **35**, 84
    type, **56**
    type parameter, **27**, 29
    user, 11, **22**
interrupt, 71, 192
interrupt handler, 71, 192
    and stack usage, 38

keyspace, 152, 157, 160, **174**

Leds interface, 80
little-endian, 43
low-power listening (radio), 216, 251
LowPowerListening interface, 216, 251

malloc, problems with, 37
Matchbox filing system, 183
McuPowerOverride interface, 218
McuPowerState interface, 219
McuSleep interface, 218
memory
    allocation, 37, 248
    buffer swap in Receive, 39
    conserving with enum, 41
    ownership, **39**, 91, 93
    sharing across components, 39, 41
Message (Java class), 115
message_t, **89**
micaz mote, **5**

mig (PC tool), 114
    and AM types, 118
    generated methods, 115
    receiving packets, 117
    sending packets, 116
module, 10, **30**
    generic, **33**, 131, 150, 164
    variable initializers, 33
mote, **4**
    micaz, 5
    power budget, 7
    Telos, 5
MoteIF (Java class), 116, 121
    and AM types, 118
    receiving packets, 117
    sending packets, 116
Mount interface, 103
multi-hop networking, 95
    collection, *see* collection (network protocol)
    dissemination, *see* dissemination (network protocol)
multiple wiring, **63**
    order of calls, 63

naming convention
    for components, 58
ncg (PC tool), 118, 176
    and #define, 119
nesC
    comparison with C and C++, 16
    compilation model, 13
nesC reference manual, 7
nesdoc, 12, 61
net.tinyos.message
    Message class, *see* Message (Java class)
    MoteIF class, *see* MoteIF (Java class)
networking
    multi-hop, 95
    serial-port, *see* serial-port networking
    single-hop, 89
norace, 229
nx_, nxle_ type prefixes, **43**

packet
    reception, 93
    sending, 90
    size, **89**
    specified with platform-independent types, 90, 112
    structure, 89
    structure (serial), 112, 113
    unreliable delivery (radio), 89
    unreliable delivery (serial), 112
packet source (PC), 116, **119**

parameterized interface, **137**, 145, 154, 156
    and default handlers, 142
    and dynamic dispatch, 140
    and unique, 146
    implementation in nesC, 139
    in modules, 140
    use in Dispatcher pattern, 167
    use in Keymap pattern, 177
    use in Keyspace pattern, 175
    use in power lock implementation, 202
    use in Service Instance pattern, 171
PC tools, 112
    Java, 112
    mig, 114
    MoteIF, *see* MoteIF (Java class)
    ncg, 118, 176
    other languages, 112
    platform-independent types, 112
    serial forwarder, **120**
permanent storage, 101
    block data, 106
    configuration data, 103
    log data, 108
    reliability, 101, 103, 110
    sampling to, 106
    volume configuration file, 102
    volumes, 102, 176
platform-independent types, **43**
platform-independent types
    accessed using mig, 114
    PC tools, 112
    use in networking, 90, 97, 112
portability, 206, 210, 221, 224
posting a task, **72**
power lock, **200**
    arbiter, 202
    configurator, 202, 204
    library, 205
    power manager, 202
    split-phase access, 200
    use in SoundLocalizer, 231
power management, 251
    for the microcontroller, 218
    in dedicated drivers, 216
    in device drivers, 215
    in shared drivers, 217
    in the radio, 216, 251
    in virtualized drivers, 217
    scheduler interaction, 218
    using power locks, 203
printf, 250

race condition, **193**, *see* data race
random number generation, 249
Read interface, 84
ReadStream interface, 87

Receive interface, 93
recursion
    avoiding, 37
    due to events signaled in commands, 76
    use tasks to avoid, 77
reliable transmission (serial), 120
Resource interface, 201
ResourceConfigure interface, 204
ResourceDefaultOwnewr interface, 203
RootControl interface, 100

sampling sensors, 84
sampling to permanent storage, 106
Send interface, 96
sending packets, 90
sensor, 84
    components, 85
    stream sampling, 87
    values, calibration, 86
sensor networks, **3**
    structure (typical), 95
sensor node, **4**
serial forwarder (PC tool), **120**
serial-port networking, 113, 241
service
    starting and stopping, 92, 97
signature (of component), **21**
single stack, 35, 37
single-hop networking, 89
singleton (component), **33**, 70
SoundLocalizer (application), 221
source code, for example applications, 9
split-phase, 6, **34**, 84
SplitControl interface, 92
stack usage, 37
StdControl interface, 97
synchronous (sync) code, **71**, 192

task, 6, 71, **72**, 192
    for avoiding recursion, 77
    posting from interrupt handlers, 199
    posting overhead, 74
    split-phase operations, 75
    timing, effects on latency and throughput, 74
Telos mote, **5**
time synchronization, 222, 225
Timer interface, 81
TinyOS
    1.x, 134, 144, 211
    API, 7, 241
    compiling, 8
    enhancement proposal (TEP), 241
    installing, 8
    overview, 5
    stack, 35, 37
    task scheduler, 72

TinyOS component
    ActiveMessageC, 94, 243
    AMReceiverC, 94, 242
    AMSenderC, 94, 242
    AMSnooperC, 242
    AMSnoopingReceiverC, 242
    BigQueueC, 248
    BitVectorC, 247
    BlockStorageC, 106, 246
    CollectionC, 101
    CollectionControlC, 243
    CollectionReceiverC, 243
    CollectionSenderC, 98, 243
    ConfigStorageC, 102, 246
    ConstantSensorC, 246
    CrcC, 250
    DemoSensorC, 246
    DisseminationC, 101, 244
    DisseminatorC, 98, 101, 244
    FcfsArbiterC, 205
    LedsC, 83, 250
    LocalTimeMilliC, 245
    LogStorageC, 108, 246
    MainC, 83, 241
    McuSleepC, 218
    NoLedsC, 250
    PoolC, 69, 248
    PrintfC, 250
    QueueC, 247
    RandomC, 249
    RandomLfsrC, 249
    RandomMlcgC, 249
    RoundArbiterC, 205
    sensors, 85, 245
    SerialActiveMessageC, 94, 243
    SerialAMReceiverC, 94, 243
    SerialAMSenderC, 94, 243
    SineSensorC, 246
    StateC, 249
    TimerMilliC, 83, 245
TinyOS interface
    Alarm, 231
    AMSend, 90
    BlockRead, 108
    BlockWrite, 106
    Boot, 81
    ConfigStorage, 103
    Counter, 226
    DisseminationUpdate, 99
    DisseminationValue, 97
    GeneralIO, 235
    I2CPacket, 236
    Leds, 80
    LowPowerListening, 216, 251
    McuPowerOverride, 218
    McuPowerState, 219

McuSleep, 218
Mount, 103
Read, 84
ReadStream, 87
Receive, 93
Resource, 201
ResourceConfigure, 204
ResourceDefaultOwnewr, 203
RootControl, 100
Send, 96
SplitControl, 92
StdControl, 97
Timer, 81
TOS_NODE_ID, 59, 91
TOSH_DATA_LENGTH, **89**
tree collection, *see* collection
type
  big-endian, 43
  little-endian, 43
  of interfaces, **56**
typedef
  in configuration, 150

unique (compile-time function), **146**, 151, 161
  use in Local Keyspace pattern, 174
  use in power lock implementation, 202
  use in Service Instance pattern, 171
unique (compile-time) function
  use in shared device drivers, 214
uniqueCount (compile-time function), **147**
  use in Local Keyspace pattern, 176
  use in power lock implementation, 202
  use in Service Instance pattern, 171

virtualized
  device, 211
  service, 147, 156
volume configuration file (for storage), 102

wiring, 12, 14, **31**
  code generated for, 51
  fan-in, **64**
  fan-out, **64**
  multiple, **63**
  omitting interfaces in, 56
  only a metaphor, 65
  use in Facade pattern, 183
  use in Keymap pattern, 177
  use in Placeholder pattern, 180